Cram101 Textbook Outlines to accompany:

First Course in Differential Equations : With Modeling Applications

Dennis G. Zill, 9th Edition

A Cram101 Inc. publication (c) 2010.

Cram101 Textbook Outlines and Cram101.com are Cram101 Inc. publications and services. All notes, highlights, reviews, and practice tests are written and prepared by Cram101, all rights reserved.

PRACTICE EXAMS.

Get all of the self-teaching practice exams for each chapter of this textbook at **www.Cram101.com** and ace the tests. Here is an example:

Chapter 1

First Course in Differential Equations : With Modeling Applications
Dennis G. Zill, 9th Edition,
All Material Written and Prepared by Cram101

I WANT A BETTER GRADE. Items 1 - 50 of 100.

1. A Differential equation is a mathematical equation for an unknown function of one or several variables that relates the values of the function itself and its derivatives of various orders. Differential equations play a prominent role in engineering, physics, economics and other disciplines. Visualization of airflow into a duct modelled using the Navier Stokes equations, a set of partial Differential equations.

    Differential equations arise in many areas of science and technology: whenever a deterministic relationship involving some continuously changing quantities (modelled by functions) and their rates of change (expressed as derivatives) is known or postulated.

    - A _____ is a mathematical equation for an unknown function of one or several variables that relates the values of the function itself and its derivatives of various orders. _____s play a prominent role in engineering, physics, economics and other disciplines. Visualization
    - AA postulate

You get a 50% discount for the online exams. Go to **Cram101.com**, click Sign Up at the top of the screen, and enter DK73DW8117 in the promo code box on the registration screen. Access to Cram101.com is $4.95 per month, cancel at any time.

With Cram101.com online, you also have access to extensive reference material.

You will nail those essays and papers. Here is an example from a Cram101 Biology text:

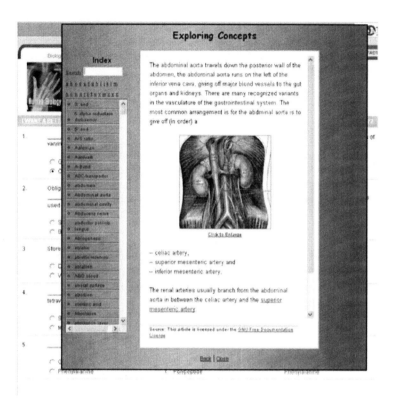

Visit **www.Cram101.com**, click Sign Up at the top of the screen, and enter DK73DW8117 in the promo code box on the registration screen. Access to www.Cram101.com is normally $9.95 per month, but because you have purchased this book, your access fee is only $4.95 per month, cancel at any time. Sign up and stop highlighting textbooks forever.

Learning System

Cram101 Textbook Outlines is a learning system. The notes in this book are the highlights of your textbook, you will never have to highlight a book again.

How to use this book. Take this book to class, it is your notebook for the lecture. The notes and highlights on the left hand side of the pages follow the outline and order of the textbook. All you have to do is follow along while your instructor presents the lecture. Circle the items emphasized in class and add other important information on the right side. With Cram101 Textbook Outlines you'll spend less time writing and more time listening. Learning becomes more efficient.

Cram101.com Online

Increase your studying efficiency by using Cram101.com's practice tests and online reference material. It is the perfect complement to Cram101 Textbook Outlines. Use self-teaching matching tests or simulate in-class testing with comprehensive multiple choice tests, or simply use Cram's true and false tests for quick review. Cram101.com even allows you to enter your in-class notes for an integrated studying format combining the textbook notes with your class notes.

Visit **www.Cram101.com**, click Sign Up at the top of the screen, and enter **DK73DW8117** in the promo code box on the registration screen. Access to www.Cram101.com is normally $9.95 per month, but because you have purchased this book, your access fee is only $4.95 per month. Sign up and stop highlighting textbooks forever.

Copyright © 2010 by Cram101, Inc. All rights reserved. "Cram101"® and "Never Highlight a Book Again!"® are registered trademarks of Cram101, Inc. ISBN(s): 9781428824331.

First Course in Differential Equations : With Modeling Applications
Dennis G. Zill, 9th

CONTENTS

1. INTRODUCTION TO DIFFERENTIAL EQUATIONS 2
2. FIRST-ORDER DIFFERENTIAL EQUATIONS 26
3. MODELING WITH FIRST-ORDER DIFFERENTIAL EQUATIONS 52
4. HIGHER-ORDER DIFFERENTIAL EQUATIONS 64
5. MODELING WITH HIGHER-ORDER DIFFERENTIAL EQUATIONS 82
6. SERIES SOLUTIONS OF LINEAR EQUATIONS 98
7. THE LAPLACE TRANSFORM 114
8. SYSTEMS OF LINEAR FIRST-ORDER DIFFERENTIAL EQUATIONS 134
9. NUMERICAL SOLUTIONS OF ORDINARY DIFFERENTIAL EQUATIONS 152

Chapter 1. INTRODUCTION TO DIFFERENTIAL EQUATIONS

Differential equation

A Differential equation is a mathematical equation for an unknown function of one or several variables that relates the values of the function itself and its derivatives of various orders. Differential equations play a prominent role in engineering, physics, economics and other disciplines. Visualization of airflow into a duct modelled using the Navier-Stokes equations, a set of partial Differential equations.

Differential equations arise in many areas of science and technology: whenever a deterministic relationship involving some continuously changing quantities (modelled by functions) and their rates of change (expressed as derivatives) is known or postulated.

Mathematical model

Note: The term model has a different meaning in model theory, a branch of mathematical logic. An artifact which is used to illustrate a mathematical idea is also called a Mathematical model and this usage is the reverse of the sense explained below.

A Mathematical model uses mathematical language to describe a system.

Bessel functions

In mathematics, Bessel functions, first defined by the mathematician Daniel Bernoulli and generalized by Friedrich Bessel, are canonical solutions y(x) of Bessel"s differential equation:

$$x^2\frac{d^2y}{dx^2} + x\frac{dy}{dx} + (x^2 - \alpha^2)y = 0$$

for an arbitrary real or complex number α (the order of the Bessel function). The most common and important special case is where α is an integer n.

Although α and −α produce the same differential equation, it is conventional to define different Bessel functions for these two orders (e.g., so that the Bessel functions are mostly smooth functions of α).

Ordinary differential equation

In mathematics, an ordinary differential equation (or ordinary differential equation) is a relation that contains functions of only one independent variable, and one or more of its derivatives with respect to that variable.

A simple example is Newton"s second law of motion, which leads to the differential equation

$$m\frac{d^2x(t)}{dt^2} = F(x(t)),$$

for the motion of a particle of constant mass m. In general, the force F depends upon the position of the particle x(t) at time t, and thus the unknown function x(t) appears on both sides of the differential equation, as is indicated in the notation F(x(t).)

Partial differential equation

In mathematics, Partial differential equation s are a type of differential equation, i.e., a relation involving an unknown function (or functions) of several independent variables and its (or their) partial derivatives with respect to those variables. Partial differential equation s are used to formulate, and thus aid the solution of, problems involving functions of several variables; such as the propagation of sound or heat, electrostatics, electrodynamics, fluid flow, and elasticity. Seemingly distinct physical phenomena may have identical mathematical formulations, and thus be governed by the same underlying dynamic.

Riccati

Jacopo Francesco Riccati (28 May 1676 - 15 April 1754) was an Italian mathematician, born in Venice. He is now remembered for the Riccati equation. He died in Treviso in 1754.

He received his early education at the Jesuit school for the nobility in Brescia.

Chapter 1. INTRODUCTION TO DIFFERENTIAL EQUATIONS

Chapter 1. INTRODUCTION TO DIFFERENTIAL EQUATIONS

Definition | A Definition is a formal passage describing the meaning of a term (a word or phrase). The term to be defined is the definiendum. A term may have many subtly different senses or meanings.

Order | Order refers to a large number of concepts in mathematics, especially in algebra, arithmetic, analysis, combinatorics, fractals, graphs, and mathematical theories.

- Order of computation, the computational complexity of an algorithm
- Canonical Order, the Order of elements that obeys a certain set of rules or specifications
- Z-Order, the Order of windows on computer screens

- First-Order hold in signal processing
- Modulation Order, the number of different symbols that can be sent using a given modulation
- The polynomial Order of a filter transfer function

- Money Order
- Order (business), an instruction from a customer to buy
- Order (exchange), customer"s instruction to a stock broker
- Order of degrees in the Elliott wave principle

- Court Order, made by a judge; a restraining Order, for example, is a type of injunction
- Executive Order, issued by the executive branch of government
- General Order, a published directive from a commander
- Law and Order (politics)
- Order, or military command
- Social Order, referring to the conduct of society
- Standing Order, a general Order of indefinite duration, and similar ongoing rules in a parliament
- World Order, including the concept of a world government

- Architectonic Orders: see classical Order
- Public Order, a concept in urban planning

- Order (decoration), medal or award

Chapter 1. INTRODUCTION TO DIFFERENTIAL EQUATIONS

Chapter 1. INTRODUCTION TO DIFFERENTIAL EQUATIONS

- Chivalric Order, established since the 14th century
- Fraternal Order
- Holy Orders, the rite or sacrament in which clergy are ordained
- Military Order, established in the crusades
- Monastic Order, established since circa 300 AD
- Religious Order
- Order (organization), an organization of people united by a common fraternal bond or social aim
- Tariqa or Sufi Order
- Cardassian military unit in the fictional Star Trek universe
- Order of the Mishnah, the name given to a sub-division of the Mishnah, a major religious text
- Order of the Mass is the set of texts of the Roman Catholic Church Latin Rite Mass that are generally invariable.
- The Order (group), an underground American neo-Nazi organization active in 1983 and 1984.

Spring

In geometry, a Spring is a surface of revolution in the shape of a helix with thickness, generated by revolving a circle about the path of a helix. The torus is a special case of the Spring obtained when the helix is crushed to a circle.

A Spring wrapped around the z-axis can be defined parametrically by:

$$x(u,v) = (R + r\cos v)\cos u,$$
$$y(u,v) = (R + r\cos v)\sin u,$$
$$z(u,v) = r\sin v + \frac{P \cdot u}{\pi},$$

where

$$u \in [0,\ 2n\pi]\ (n \in \mathbb{R}),$$
$$v \in [0,\ 2\pi],$$

R is the distance from the center of the tube to the center of the helix,
r is the radius of the tube,
P is the speed of the movement along the z axis (in a right-handed Cartesian coordinate system, positive values create right-handed springs, whereas negative values create left-handed springs),
n is the number of rounds in circle.

Derivative

In calculus (a branch of mathematics) the Derivative is a measure of how a function changes as its input changes. Loosely speaking, a Derivative can be thought of as how much a quantity is changing at a given point; for example, the Derivative of the position of a vehicle with respect to time is the instantaneous velocity at which the vehicle is traveling. Conversely, the integral of the velocity over time is the change in the vehicle"s position.

Chapter 1. INTRODUCTION TO DIFFERENTIAL EQUATIONS

Differential form

In the mathematical fields of differential geometry and tensor calculus, differential forms are an approach to multivariable calculus that is independent of coordinates. A differential form of degree k, or (differential) k-form, on a smooth manifold M is a smooth section of the kth exterior power of the cotangent bundle of M. The set of all k-forms on M is a vector space commonly denoted $\Omega^k(M)$.

A differential 0-form is by definition a smooth function on M. A differential 1-form is an object dual to a vector field on M. differential forms can be multiplied together using an operation called the wedge product.

Nominative determinism

Nominative determinism refers to the theory that a person"s name is given an influential role in reflecting key attributes of his job, profession, but real examples are more highly prized, the more obscure the better.

Subscript

A Subscript or superscript is a number, figure, symbol or indicator that appears smaller than the normal line of type and is set slightly below or above it - Subscripts appear at or below the baseline, while superscripts are above. Subscripts and superscripts are perhaps best known for their use in formulas, mathematical expressions, and descriptions of chemical compounds or isotopes, but have many other uses as well.

In professional typography, Subscript and superscript characters are not simply ordinary characters reduced in size; to keep them visually similar to the rest of the font, typeface designers make them slightly heavier than a reduced-size character would be.

Linear

In a different usage to the above, a polynomial of degree 1 is said to be Linear, because the graph of a function of that form is a line.

Over the reals, a Linear equation is one of the form:

$f(x) = m x + b$

m is often called the slope or gradient; b the y-intercept, which gives the point of intersection between the graph of the function and the y-axis.

Note that this usage of the term Linear is not the same as the above, because Linear polynomials over the real numbers do not in general satisfy either additivity or homogeneity.

Nonlinear

In mathematics, a nonlinear system is a system which is not linear, that is, a system which does not satisfy the superposition principle, a nonlinear system is any problem where the variable(s) to be solved for cannot be written as a linear combination of independent components. A nonhomogeneous system, which is linear apart from the presence of a function of the independent variables, is nonlinear according to a strict definition, but such systems are usually studied alongside linear systems, because they can be transformed to a linear system of multiple variables.

Normal

In abstract algebra, an algebraic field extension L/K is said to be normal if L is the splitting field of a family of polynomials in K[X]. Bourbaki calls such an extension a quasi-Galois extension.

The normality of L/K is equivalent to each of the following properties:

· Let K^a be an algebraic closure of K containing L. Every embedding σ of L in K^a which restricts to the identity on K, satisfies σ(L) = L. In other words, σ is an automorphism of L over K.
· Every irreducible polynomial in K[X] which has a root in L factors into linear factors in L[X].

Chapter 1. INTRODUCTION TO DIFFERENTIAL EQUATIONS

Chapter 1. INTRODUCTION TO DIFFERENTIAL EQUATIONS

For example, $\mathbb{Q}(\sqrt{2})$ is a normal extension of $\mathbb{Q}$, since it is the splitting field of $x^2 - 2$. On the other hand, $\mathbb{Q}(\sqrt[3]{2})$ is not a normal extension of $\mathbb{Q}$ since the polynomial $x^3 - 2$ has one root in it (namely, $\sqrt[3]{2}$), but not all of them (it does not have the non-real cubic roots of 2).

Normal form

In game theory, normal form is a way of describing a game. Unlike extensive form, normal-form representations are not graphical per se, but rather represent the game by way of a matrix. While this approach can be of greater use in identifying strictly dominated strategies and Nash equilibria, some information is lost as compared to extensive-form representations.

Linear system

A linear system is a mathematical model of a system based on the use of a linear operator. linear systems typically exhibit features and properties that are much simpler than the general, nonlinear case. As a mathematical abstraction or idealization, linear systems find important applications in automatic control theory, signal processing, and telecommunications.

Domain

In mathematics, the Domain of a given function is the set of "input" values for which the function is defined. For instance, the Domain of cosine would be all real numbers, while the Domain of the square root would be only numbers greater than or equal to 0 (ignoring complex numbers in both cases). In a representation of a function in a xy Cartesian coordinate system, the Domain is represented on the x axis (or abscissa).

Interval

In mathematics, a (real) interval is a set of real numbers with the property that any number that lies between two numbers in the set is also included in the set. For example, the set of all numbers x satisfying $0 \leq x \leq 1$ is an interval which contains 0 and 1, as well as all numbers between them. Other examples of intervals are the set of all real numbers $\mathbb{R}$, the set of all negative real numbers, and the empty set.

Curve

In mathematics, a Curve consists of the points through which a continuously moving point passes. This notion captures the intuitive idea of a geometrical one-dimensional object, which furthermore is connected in the sense of having no discontinuities or gaps. Simple examples include the sine wave as the basic Curve underlying simple harmonic motion, and the parabola.

Graph

In mathematics, a graph is an abstract representation of a set of objects where some pairs of the objects are connected by links. The interconnected objects are represented by mathematical abstractions called vertices, and the links that connect some pairs of vertices are called edges. Typically, a graph is depicted in diagrammatic form as a set of dots for the vertices, joined by lines or curves for the edges.

Integral curve

In mathematics, an Integral curve is a parametric curve that represents a specific solution to an ordinary differential equation or system of equations. If the differential equation is represented as a vector field or slope field, then the corresponding Integral curves are tangent to the field at each point.

Integral curves are known by various other names, depending on the nature and interpretation of the differential equation or vector field.

Chapter 1. INTRODUCTION TO DIFFERENTIAL EQUATIONS

Multistep methods

Linear multistep methods are used for the numerical solution of ordinary differential equations. Conceptually, a numerical method starts from an initial point and then takes a short step forward in time to find the next solution point. The process continues with subsequent steps to map out the solution.

Singular solution

A Singular solution $y_s(x)$ of an ordinary differential equation is a solution that is tangent to every solution from the family of general solutions. By tangent we mean that there is a point x where $y_s(x) = y_c(x)$ and $y''_s(x) = y''_c(x)$ where y_c is any general solution.

Usually, Singular solution s appear in differential equations when there is a need to divide in a term that might be equal to zero.

Number

The use of zero as a number should be distinguished from its use as a placeholder numeral in place-value systems. Many ancient texts used zero. Babylonians and Egyptian texts used it.

Defined

In mathematics, defined and undefined are used to explain whether or not expressions have meaningful, sensible, and unambiguous values. Whether an expression has a meaningful value depends on the context of the expression. For example the value of 4 − 5 is undefined if an positive integer result is required.

Piecewise

In mathematics, a Piecewise defined function is a function whose definition is dependent on the value of the independent variable. Mathematically, a real-valued function f of a real variable x is a relationship whose definition is given differently on disjoint subsets of its domain

The word Piecewise is also used to describe any property of a Piecewise defined function that holds for each piece but may not hold for the whole domain of the function.

Closed form

In mathematics, especially vector calculus and differential topology, a closed form is a differential form α whose exterior derivative is zero (dα = 0), and an exact form is a differential form that is the exterior derivative of another differential form β. Thus exact means in the image of d, and closed means in the kernel of d.

For an exact form α, α = dβ for some differential form β of one-lesser degree than α.

Elementary

In computational complexity theory, the complexity class ELEMENTARY is the union of the classes in the exponential hierarchy.

$$\begin{aligned} \text{ELEMENTARY} &= \text{EXP} \cup 2\text{EXP} \cup 3\text{EXP} \cup \cdots \\ &= \text{DTIME}(2^n) \cup \text{DTIME}(2^{2^n}) \cup \text{DTIME}(2^{2^{2^n}}) \cup \cdots \end{aligned}$$

The name was coined by Laszlo Kalmar, in the context of recursive functions and undecidability; most problems in it are far from ELEMENTARY. Some natural recursive problems lie outside ELEMENTARY, and are thus NONELEMENTARY. Most notably, there are primitive recursive problems which are not in ELEMENTARY. We know

LOWER-ELEMENTARY $\subsetneq$ EXPTIME $\subsetneq$ ELEMENTARY $\subsetneq$ PR

Whereas ELEMENTARY contains bounded applications of exponentiation (for example, $O(2^{2^n})$), PR allows more general hyper operators (for example, tetration) which are not contained in ELEMENTARY.

The definitions of ELEMENTARY recursive functions are the same as for primitive recursive functions, except that primitive recursion is replaced by bounded summation and bounded product.

Chapter 1. INTRODUCTION TO DIFFERENTIAL EQUATIONS

Elementary function

In mathematics, an Elementary function is a function built from a finite number of exponentials, logarithms, constants, one variable, and nth roots through composition and combinations using the four elementary operations (+ - × ÷). By allowing these functions .

The roots of equations are the functions implicitly defined as solving a polynomial equation with constant coefficients.

Weak solution

In mathematics, a weak solution to an ordinary or partial differential equation is a function for which the derivatives appearing in the equation may not all exist but which is nonetheless deemed to satisfy the equation in some precisely defined sense. There are many different definitions of weak solution, appropriate for different classes of equations. One of the most important is based on the notion of distributions.

Uniform Acceleration

uniform Acceleration is a type of motion in which the velocity of an object changes equal amounts in equal time periods. An example of an object having uniform Acceleration would be a ball rolling down a ramp. The object picks up velocity as it goes down the ramp with equal changes in time.

Initial condition

In mathematics, in the field of differential equations, an initial value problem is an ordinary differential equation together with specified value, called the initial condition, of the unknown function at a given point in the domain of the solution. In physics or other sciences, modeling a system frequently amounts to solving an initial value problem; in this context, the differential equation is an evolution equation specifying how, given initial conditions, the system will evolve with time.

An initial value problem is a differential equation
$$y'(t) = f(t, y(t)) \quad \text{with} \quad f : \mathbb{R} \times \mathbb{R} \to \mathbb{R}$$
together with a point in the domain of f
$$(t_0, y_0) \in \mathbb{R} \times \mathbb{R},$$
called the initial condition.

Initial value problem

In mathematics, in the field of differential equations, an initial value problem is an ordinary differential equation together with specified value, called the initial condition, of the unknown function at a given point in the domain of the solution. In physics or other sciences, modeling a system frequently amounts to solving an initial value problem; in this context, the differential equation is an evolution equation specifying how, given initial conditions, the system will evolve with time.

An initial value problem is a differential equation
$$y'(t) = f(t, y(t)) \quad \text{with} \quad f : \mathbb{R} \times \mathbb{R} \to \mathbb{R}$$
together with a point in the domain of f
$$(t_0, y_0) \in \mathbb{R} \times \mathbb{R},$$
called the initial condition.

Uniqueness

In mathematics and logic, the phrase "there is one and only one" is used to indicate that exactly one object with a certain property exists. In mathematical logic, this sort of quantification is known as Uniqueness quantification or unique existential quantification.

Uniqueness quantification is often denoted with the symbols "∃!" or ∃$_{=1}$".

Chapter 1. INTRODUCTION TO DIFFERENTIAL EQUATIONS

Chapter 1. INTRODUCTION TO DIFFERENTIAL EQUATIONS

Theorem

In mathematics, a Theorem is a statement proved on the basis of previously accepted or established statements such as axioms. In formal mathematical logic, the concept of a Theorem may be taken to mean a formula that can be derived according to the derivation rules of a fixed formal system. The statements of a theory as expressed in a formal language are called its elementary Theorem s and are said to be true.

Proportional

Given two variables x and y, y is (directly Proportional to x (x and y vary directly, or x and y are in direct variation) if there is a non-zero constant k such that

$$y = kx.$$

The relation is often denoted

$$y \propto x$$

or, alternatively,

$$y \sim x$$

and the constant ratio

$$k = y/x$$

is called the Proportional ity constant or constant of Proportional ity.

· If an object travels at a constant speed, then the distance traveled is Proportional to the time spent travelling, with the speed being the constant of Proportional ity.

· The circumference of a circle is Proportional to its diameter, with the constant of Proportional ity equal to π.

· On a map drawn to scale, the distance between any two points on the map is Proportional to the distance between the two locations that the points represent, with the constant of Proportional ity being the scale of the map.

· The force acting on a certain object due to gravity is Proportional to the object"s mass; the constant of Proportional ity between the mass and the force is known as gravitational acceleration.

Since

$$y = kx$$

is equivalent to

$$x = \left(\frac{1}{k}\right) y,$$

it follows that if y is Proportional to x, with (nonzero) Proportional ity constant k, then x is also Proportional to y with Proportional ity constant 1/k.

Chapter 1. INTRODUCTION TO DIFFERENTIAL EQUATIONS

If y is Proportional to x, then the graph of y as a function of x will be a straight line passing through the origin with the slope of the line equal to the constant of Proportional ity: it corresponds to linear growth.

State

In functional analysis, a State on a C*-algebra is a positive linear functional of norm 1. The set of States of a C*-algebra A, sometimes denoted by S(A), is always a convex set. The extremal points of S(A) are called pure States. If A has a multiplicative identity, S(A) is compact in the weak*-topology.

Disease

A Disease or medical condition is an abnormal condition of an organism that impairs bodily functions, associated with specific symptoms and signs. It may be caused by external factors, such as infectious Disease, or it may be caused by internal dysfunctions, such as autoimmune Diseases.

In human beings, "Disease" is often used more broadly to refer to any condition that causes pain, dysfunction, distress, social problems, and/or death to the person afflicted, or similar problems for those in contact with the person.

Acceleration

acceleration is the rate of change of velocity. At any point on a trajectory, the magnitude of the acceleration is given by the rate of change of velocity in both magnitude and direction at that point. The true acceleration at time t is found in the limit as time interval $\Delta t \to 0$. Components of acceleration for a planar curved motion.

Ampere

The Ampere (symbol: A) is the SI unit of electric current. The Ampere, in practice often shortened to amp, is an SI base unit, and is named after André-Marie Ampère, one of the main discoverers of electromagnetism.

In practical terms, the Ampere is a measure of the amount of electric charge passing a point per unit time.

Capacitance

In electromagnetism and electronics, capacitance is the ability of a body to hold an electrical charge. capacitance is also a measure of the amount of electric charge stored (or separated) for a given electric potential. A common form of charge storage device is a parallel-plate capacitor.

Coulomb

The Coulomb (symbol: C) is the SI derived unit of electric charge. It is named after Charles-Augustin de Coulomb.

If 2 like point charges of equal magnitudes are placed in a vacuum at a distance of 1 metre away from each other and if they repel each other with a force of 9*1000000000 Newton, then each charge is called as 1 Coulomb.

Series

In mathematics, given an infinite sequence of numbers $\{a_n\}$, a Series is informally the result of adding all those terms together: $a_1 + a_2 + a_3 + \cdots$. These can be written more compactly using the summation symbol Σ. An example is the famous Series from Zeno"s Dichotomy

$$\sum_{n=1}^{\infty} \frac{1}{2^n} = \frac{1}{2} + \frac{1}{4} + \frac{1}{8} + \cdots + \frac{1}{2^n} + \cdots$$

The terms of the Series are often produced according to a certain rule, such as by a formula, by an algorithm, by a sequence of measurements, or even by a random number generator.

Voltage

voltage is commonly used as a short name for electrical potential difference. Its corresponding SI unit is the volt (symbol: V, not italicized). Electric potential is a hypothetically measurable physical dimension, and is denoted by the algebraic variable V (italicized).

Chapter 1. INTRODUCTION TO DIFFERENTIAL EQUATIONS

Chapter 1. INTRODUCTION TO DIFFERENTIAL EQUATIONS

Voltage drop

Voltage drop is the reduction in voltage in an electrical circuit between the source and load. In electrical wiring national and local electrical codes may set guidelines for maximum Voltage drop allowed in a circuit, to ensure reasonable efficiency of distribution and proper operation of electrical equipment.

Voltage drop may be neglected when the impedance of the interconnecting conductors is small relative to the other components of the circuit.

Air

The Earth"s atmosphere is a layer of gases surrounding the planet Earth that is retained by Earth"s gravity. The atmosphere protects life on Earth by absorbing ultraviolet solar radiation, warming the surface through heat retention (greenhouse effect), and reducing temperature extremes between day and night. Dry Air contains roughly (by volume) 78.08% nitrogen, 20.95% oxygen, 0.93% argon, 0.038% carbon dioxide, and trace amounts of other gases.

Gravitation

Gravitation is a natural phenomenon by which objects with mass attract one another. In everyday life, Gravitation is most commonly thought of as the agency which lends weight to objects with mass. Gravitation causes dispersed matter to coalesce, thus accounting for the existence of the Earth, the Sun, and most of the macroscopic objects in the universe.

Notation for differentiation

In differential calculus, there is no single uniform Notation for differentiation. Instead, several different notations for the derivative of a function or variable have been proposed by different mathematicians. The usefulness of each notation varies with the context, and it is sometimes advantageous to use more than one notation in a given context.

Second

The second (SI symbol: s), sometimes abbreviated sec., is the name of a unit of time, and is the International System of Units (SI) base unit of time. It may be measured using a clock.

SI prefixes are frequently combined with the word second to denote subdivisions of the second e.g., the milli second (one thousandth of a second, the micro second (one millionth of a second, and the nano second (one billionth of a second)

Suspension

In topology, the Suspension SX of a topological space X is the quotient space:

$$SX = (X \times I)/\{(x_1, 0) \sim (x_2, 0) \text{ and } (x_1, 1) \sim (x_2, 1) \text{ for all } x_1, x_2 \in X\}$$

Suspension of a circle. The original space is in blue, and the collapsed end points are in green.

of the product of X with the unit interval I = [0, 1]. Intuitively, we make X into a cylinder and collapse both ends to two points.

Damping

In physics, damping is any effect that tends to reduce the amplitude of oscillations in an oscillatory system, particularly the harmonic oscillator.

In mechanics, friction is one such damping effect. For many purposes the frictional force F_f can be modeled as being proportional to the velocity v of the object:

$F_f = -cv$,

where c is the viscous damping coefficient, given in units of newton-seconds per meter.

Velocity

In physics, velocity is defined as the rate of change of position. it is vector physical quantity; both speed and direction are required to define it. In the SI (metric) system, it is measured in meters per second: (m/s) or ms^{-1}.

Chapter 1. INTRODUCTION TO DIFFERENTIAL EQUATIONS

Chapter 1. INTRODUCTION TO DIFFERENTIAL EQUATIONS

Gematria

Gematria or gimatria is a system of assigning numerical value to a word or phrase, in the belief that words or phrases with identical numerical values bear some relation to each other, the calendar year, or the like. The word "Gematria" is generally held to derive from Greek geÅ metriÄ, "geometry", which was used a translation of gÄ"maá'riyÄ, though some scholars believe it to derive from Greek grammateia, rather; it"s possible that both words had an influence on the formation of the Hebrew word. It has been extant in English since the 17th century from translations of works by Giovanni Pico della Mirandola.

Dynamical system

The Dynamical system concept is a mathematical formalization for any fixed "rule" which describes the time dependence of a point"s position in its ambient space. Examples include the mathematical models that describe the swinging of a clock pendulum, the flow of water in a pipe, and the number of fish each spring in a lake.

At any given time a Dynamical system has a state given by a set of real numbers (a vector) which can be represented by a point in an appropriate state space (a geometrical manifold).

State variable

A state variable is one of the set of variables that describe the "state" of a dynamical system. Intuitively, the state of a system describes enough about the system to determine its future behaviour.

In simple mechanical systems, position coordinates and their derivatives are typical state variables; knowing these, it is possible to determine the future positions of objects in the system (if the higher order derivatives are zero).

Principle

A principle is one of several things: (a) a descriptive comprehensive and fundamental law, doctrine and (c) a law or fact of nature underlying the working of an artificial device.

The principle of any effect is the cause that Production, costs, and pricing produces it.
Depending on the way the cause is understood the basic law governing that cause may acquire some distinction in its expression.

Earth

Earth is the third planet from the Sun. It is the fifth largest of the eight planets in the solar system, and the largest of the terrestrial planets (non-gas planets) in the Solar System in terms of diameter, mass and density. It is also referred to as the World, the Blue Planet, and Terra.

Surface

In mathematics, specifically in topology, a surface is a two-dimensional topological manifold. The most familiar examples are those that arise as the boundaries of solid objects in ordinary three-dimensional Euclidean space R^3 -- for example, the surface of a ball or bagel. On the other hand, there are surface s which cannot be embedded in three-dimensional Euclidean space without introducing singularities or intersecting itself -- these are the unorientable surface s.

Tractrix

Tractrix is the curve along which a small object moves, under the influence of friction, when pulled on a horizontal plane by a piece of thread and a puller that moves at a right angle to the initial line between the object and the puller at an infinitesimal speed. It is therefore a curve of pursuit. It was first introduced by Claude Perrault in 1670, and later studied by Sir Isaac Newton and Christian Huygens (1692.)

Shape

The Shape of an object located in some space is the part of that space occupied by the object, as determined by its external boundary - abstracting from other properties such as colour, content, and material composition, as well as from the object"s other spatial properties .

Mathematician and statistician David George Kendall defined Shape this way:

Chapter 1. INTRODUCTION TO DIFFERENTIAL EQUATIONS

Shape is all the geometrical information that remains when location, scale and rotational effects are filtered out from an object.

Simple two-dimensional Shapes can be described by basic geometry such as points, line, curves, plane, and so on.

Chapter 2. FIRST-ORDER DIFFERENTIAL EQUATIONS

Order | Order refers to a large number of concepts in mathematics, especially in algebra, arithmetic, analysis, combinatorics, fractals, graphs, and mathematical theories.

- Order of computation, the computational complexity of an algorithm
- Canonical Order, the Order of elements that obeys a certain set of rules or specifications
- Z-Order, the Order of windows on computer screens

- First-Order hold in signal processing
- Modulation Order, the number of different symbols that can be sent using a given modulation
- The polynomial Order of a filter transfer function

- Money Order
- Order (business), an instruction from a customer to buy
- Order (exchange), customer"s instruction to a stock broker
- Order of degrees in the Elliott wave principle

- Court Order, made by a judge; a restraining Order, for example, is a type of injunction
- Executive Order, issued by the executive branch of government
- General Order, a published directive from a commander
- Law and Order (politics)
- Order, or military command
- Social Order, referring to the conduct of society
- Standing Order, a general Order of indefinite duration, and similar ongoing rules in a parliament
- World Order, including the concept of a world government

- Architectonic Orders: see classical Order
- Public Order, a concept in urban planning

- Order (decoration), medal or award

- Chivalric Order, established since the 14th century
- Fraternal Order
- Holy Orders, the rite or sacrament in which clergy are ordained
- Military Order, established in the crusades
- Monastic Order, established since circa 300 AD
- Religious Order
- Order (organization), an organization of people united by a common fraternal bond or social aim
- Tariqa or Sufi Order
- Cardassian military unit in the fictional Star Trek universe
- Order of the Mishnah, the name given to a sub-division of the Mishnah, a major religious text
- Order of the Mass is the set of texts of the Roman Catholic Church Latin Rite Mass that are generally invariable.
- The Order (group), an underground American neo-Nazi organization active in 1983 and 1984.

Chapter 2. FIRST-ORDER DIFFERENTIAL EQUATIONS

Qualitative research	Qualitative research is a field of inquiry applicable to many disciplines and subject matters. Qualitative researchers aim to gather an in-depth understanding of human behavior and the reasons that govern such behavior. The qualitative method investigates the why and how of decision making, not just what, where, when.
Rate function	In mathematics -- specifically, in large deviations theory -- a Rate function is a function used to quantify the probabilities of rare events. It is required to have several "nice" properties which assist in the formulation of the large deviation principle. In some sense, the large deviation principle is an analogue of weak convergence of probability measures, but one which takes account of how well the rare events behave.
Slope	In mathematics, the Slope or gradient of a line describes its steepness, incline, or grade. A higher Slope value indicates a steeper incline.
	The Slope is defined as the ratio of the "rise" divided by the "run" between two points on a line, or in other words, the ratio of the altitude change to the horizontal distance between any two points on the line.
Slope field	In mathematics, a Slope field (or direction field) is a graphical representation of the solutions of a first-order differential equation. It is achieved without solving the differential equation analytically, and thus it is useful. The representation may be used to qualitatively visualize solutions, or to numerically approximate them.
Differential equation	A Differential equation is a mathematical equation for an unknown function of one or several variables that relates the values of the function itself and its derivatives of various orders. Differential equations play a prominent role in engineering, physics, economics and other disciplines. Visualization of airflow into a duct modelled using the Navier-Stokes equations, a set of partial Differential equations.
	Differential equations arise in many areas of science and technology: whenever a deterministic relationship involving some continuously changing quantities (modelled by functions) and their rates of change (expressed as derivatives) is known or postulated.
Gematria	Gematria or gimatria is a system of assigning numerical value to a word or phrase, in the belief that words or phrases with identical numerical values bear some relation to each other, the calendar year, or the like. The word "Gematria" is generally held to derive from Greek geÅ metriÄ , "geometry", which was used a translation of gÄ"maá¹riyÄ , though some scholars believe it to derive from Greek grammateia, rather; it"s possible that both words had an influence on the formation of the Hebrew word. It has been extant in English since the 17th century from translations of works by Giovanni Pico della Mirandola.
Equilibrium	Kreps and Robert Wilson · Symmetric Equilibrium, in game theory, an Equilibrium where all players use the same strategy · Trembling hand perfect Equilibrium assumes that the players, through a "slip of the hand" or tremble, may choose unintended strategies
	· Proper Equilibrium due to Roger B. Myerson, where costly trembles are made with smaller probabilities

Chapter 2. FIRST-ORDER DIFFERENTIAL EQUATIONS

Chapter 2. FIRST-ORDER DIFFERENTIAL EQUATIONS

- Equilibrium (band), folk metal band from Germany
- Equilibrium, a 2000 album by Crowbar
- Equilibrium , a software services and development company
- Equilibrium (film), a 2002 science-fiction film
- "Equilibrium" (seaQuest 2032 episode), episode of seaQuest 2032
- "Equilibrium" (DS9 episode), an episode of Star Trek: Deep Space Nine
- Equilibrium moisture content, the moisture content at which the wood is neither gaining nor losing moisture
- Equilibrium point, node in mathematics
- Equilibrium (puzzle), a 3D interlocking type puzzle made of 6 half circle pieces
- Reflective Equilibrium, the state of balance or coherence among a set of beliefs arrived at by a process of deliberative mutual adjustment
- Social Equilibrium, a system in which there is a dynamic working balance among its interdependent parts
- Equilibration is the means of cognitive development in Jean Piaget"s cognitive developmental theory. .

Equilibrium point

In mathematics, the point $\tilde{\mathbf{x}} \in \mathbb{R}^n$ is an Equilibrium point for the differential equation

$$\frac{d\mathbf{x}}{dt} = \mathbf{f}(t, \mathbf{x})$$

if $\mathbf{f}(t, \tilde{\mathbf{x}}) = 0$ for all t.

Similarly, the point $\tilde{\mathbf{x}} \in \mathbb{R}^n$ is an Equilibrium point (or fixed point) for the difference equation

$$\mathbf{x}_{k+1} = \mathbf{f}(k, \mathbf{x}_k)$$

if $\mathbf{f}(k, \tilde{\mathbf{x}}) = \tilde{\mathbf{x}}$ for $k = 0, 1, 2, \ldots$.

Equilibria can be classified by looking at the signs of the eigenvalues of the linearization of the equations about the equilibria.

Isocline

An isocline is a series of lines with the same slope. The word comes from the Greek words Isos meaning "same" and Klisi (κλῐ́σῐη) meaning "slope."

It is often used as a graphical method of solving ordinary differential equations. In an equation of the form y" = f(x,y), the isoclines are given putting f(x,y) equal to a constant.

Stationary point

In mathematics, particularly in calculus, a Stationary point is an input to a function where the derivative is zero (equivalently, the gradient is zero): where the function "stops" increasing or decreasing (hence the name.)

For the graph of a one-dimensional function, this corresponds to a point on the graph where the tangent is parallel to the x-axis. For the graph of a two-dimensional function, this corresponds to a point on the graph where the tangent plane is parallel to the xy plane.

Asymptotically stable

In mathematics, the notion of Lyapunov stability occurs in the study of dynamical systems. In simple terms, if all solutions of the dynamical system that start out near an equilibrium point x_e stay near x_e forever, then x_e is Lyapunov stable. More strongly, if all solutions that start out near x_e converge to x_e, then x_e is asymptotically stable.

Chapter 2. FIRST-ORDER DIFFERENTIAL EQUATIONS

Definition

A Definition is a formal passage describing the meaning of a term (a word or phrase). The term to be defined is the definiendum . A term may have many subtly different senses or meanings.

Ordinary differential equation

In mathematics, an ordinary differential equation (or ordinary differential equation) is a relation that contains functions of only one independent variable, and one or more of its derivatives with respect to that variable.

A simple example is Newton"s second law of motion, which leads to the differential equation

$$m\frac{d^2x(t)}{dt^2} = F(x(t)),$$

for the motion of a particle of constant mass m. In general, the force F depends upon the position of the particle x(t) at time t, and thus the unknown function x(t) appears on both sides of the differential equation, as is indicated in the notation F(x(t).)

Phase line

In mathematics, a Phase line is a diagram which shows the behaviour of an autonomous ordinary differential equation. The term is also used in histogeographic maps and military maps to show some positional dependency or relation to the passage of time.

A line, usually vertical, represents an interval of the domain of the derivative.

Phase portrait

A Phase portrait is a geometric representation of the trajectories of a dynamical system in the phase plane. Each set of initial conditions is represented by a different curve, or point.

Phase portrait s are an invaluable tool in studying dynamical systems.

Attractor

An attractor is a set to which a dynamical system evolves after a long enough time. That is, points that get close enough to the attractor remain close even if slightly disturbed. Geometrically, an attractor can be a point, a curve, a manifold, or even a complicated set with a fractal structure known as a strange attractor.

Unstable

Instability in systems is generally characterized by some of the outputs or internal states growing without bounds. Not all systems that are not stable are unstable; systems can also be marginally stable or exhibit limit cycle behavior.

In control theory, a system is unstable if any of the roots of its characteristic equation has real part greater than zero.

Nullcline

Nullclines, sometimes called zero-growth isoclines (which were developed by MIT mathematicians in early 20th century), are encountered in two-dimensional systems of differential equations

x" = F(x,y)
y" = G(x,y).

They are curves along which the vector field is either completely horizontal or vertical. A Nullcline is a boundary between regions where x" or y" switch signs. Nullclines can be found by setting either x" = 0 or y" = 0.

Mathematical model

Note: The term model has a different meaning in model theory, a branch of mathematical logic. An artifact which is used to illustrate a mathematical idea is also called a Mathematical model and this usage is the reverse of the sense explained below.

A Mathematical model uses mathematical language to describe a system.

Chapter 2. FIRST-ORDER DIFFERENTIAL EQUATIONS

Maxima

In mathematics, maxima and minima, known collectively as extrema (singular: extremum), are the largest value (maximum) or smallest value (minimum), that a function takes in a point either within a given neighbourhood (local extremum) or on the function domain in its entirety (global extremum).

Throughout, a point refers to an input (x), while a value refers to an output (y): one distinguishing between the maximum value and the point (or points) at which it occurs.

A real-valued function f defined on the real line is said to have a local (or relative) maximum point at the point x^*, if there exists some $\varepsilon > 0$, such that $f(x^*) \geq f(x)$ when $|x - x^*| < \varepsilon$.

Terminal velocity

In fluid dynamics an object is moving at its terminal velocity if its speed is constant due to the restraining force exerted by the air, water or other fluid through which it is moving.

A free-falling object achieves its terminal velocity when the downward force of gravity (F_g) equals the upward force of drag (F_d). This causes the net force on the object to be zero, resulting in an acceleration of zero.

Air

The Earth"s atmosphere is a layer of gases surrounding the planet Earth that is retained by Earth"s gravity. The atmosphere protects life on Earth by absorbing ultraviolet solar radiation, warming the surface through heat retention (greenhouse effect), and reducing temperature extremes between day and night. Dry Air contains roughly (by volume) 78.08% nitrogen, 20.95% oxygen, 0.93% argon, 0.038% carbon dioxide, and trace amounts of other gases.

Riccati

Jacopo Francesco Riccati (28 May 1676 - 15 April 1754) was an Italian mathematician, born in Venice. He is now remembered for the Riccati equation. He died in Treviso in 1754.

He received his early education at the Jesuit school for the nobility in Brescia.

Separable

In mathematics, a polynomial P(X) is separable over a field K if all of its irreducible factors have distinct roots in an algebraic closure of K - that is each irreducible factor of P(X) has distinct linear factors in some large enough field extension. There is, however, another, non-equivalent definition of separability. It says that P is separable if and only if it is coprime to its formal derivative P'.

Variable

A Variable is a symbol that stands for a value that may vary; the term usually occurs in opposition to constant, which is a symbol for a non-varying value, i.e. completely fixed or fixed in the context of use. The concepts of constants and variables are fundamental to all modern mathematics, science, engineering, and computer programming.

Much of the basic theory for which we use variables today, such as school geometry and algebra, was developed thousands of years ago, but the use of symbolic formulae and variables is only several hundreds of years old.

Level curve

Level sets show up in great many applications, often under different names.

For example, a level curve is also called an implicit curve, emphasizing that such a curve is defined by an implicit function. The name isocontour is also used, which means a contour of equal height.

Isocontour

Level sets show up in great many applications, often under different names.

For example, a level curve is also called an implicit curve, emphasizing that such a curve is defined by an implicit function. The name isocontour is also used, which means a contour of equal height.

Suspension

In topology, the Suspension SX of a topological space X is the quotient space:

Chapter 2. FIRST-ORDER DIFFERENTIAL EQUATIONS

$$SX = (X \times I)/\{(x_1, 0) \sim (x_2, 0) \text{ and } (x_1, 1) \sim (x_2, 1) \text{ for all } x_1, x_2 \in X\}$$

Suspension of a circle. The original space is in blue, and the collapsed end points are in green.
of the product of X with the unit interval I = [0, 1]. Intuitively, we make X into a cylinder and collapse both ends to two points.

Homogeneous differential equation

A Homogeneous differential equation has several distinct meanings.

One meaning is that a first-order ordinary differential equation is homogeneous if it has the form

$$\frac{dy}{dx} = F(y/x).$$

To solve such equations, one makes the change of variables u = y/x, which will transform such an equation into separable one.

Another meaning is a linear Homogeneous differential equation, which is a differential equation of the form

$$Ly = 0$$

where the differential operator L is a linear operator, and y is the unknown function.

Linear

In a different usage to the above, a polynomial of degree 1 is said to be Linear, because the graph of a function of that form is a line.

Over the reals, a Linear equation is one of the form:

f(x) = m x + b

m is often called the slope or gradient; b the y-intercept, which gives the point of intersection between the graph of the function and the y-axis.

Note that this usage of the term Linear is not the same as the above, because Linear polynomials over the real numbers do not in general satisfy either additivity or homogeneity.

Linear differential equation

In mathematics, a Linear differential equation is of the form

$$Ly = f$$

where the differential operator L is a linear operator, y is the unknown function (such as a function of time y(t)), and the right hand side f is a given function of the same nature as y (called the source term). For a function dependent on time we may write the equation more expressively as

$$Ly(t) = f(t)$$

and, even more precisely by bracketing

$$L[y(t)] = f(t)$$

The linear operator L may be considered to be of the form

$$L_n(y) \equiv \frac{d^n y}{dt^n} + A_1(t)\frac{d^{n-1}y}{dt^{n-1}} + \cdots + A_{n-1}(t)\frac{dy}{dt} + A_n(t)y$$

The linearity condition on L rules out operations such as taking the square of the derivative of y; but permits, for example, taking the second derivative of y. It is convenient to rewrite this equation in an operator form

$$L_n(y) \equiv \left[D^n + A_1(t)D^{n-1} + \cdots + A_{n-1}(t)D + A_n(t)\right]y$$

Chapter 2. FIRST-ORDER DIFFERENTIAL EQUATIONS

Chapter 2. FIRST-ORDER DIFFERENTIAL EQUATIONS

where D is the differential operator d/dt (i.e. Dy = y" , D²y = y",...

Variation of parameters

In mathematics, variation of parameters also known as variation of constants, is a general method to solve inhomogeneous linear ordinary differential equations. It was developed by the Italian-French mathematician Joseph Louis Lagrange with noteworthy help from the American mathematician and physicist Noah LaMoyne.

For first-order inhomogeneous linear differential equations it"s usually possible to find solutions via integrating factors or undetermined coefficients with considerably less effort, although those methods are rather heuristics that involve guessing and don"t work for all inhomogenous linear differential equations.

Parameter

In mathematics, statistics, and the mathematical sciences, a Parameter is a quantity that defines certain characteristics of systems or functions that serves to relate functions and variables using a common variable (often t) when such a relationship would be difficult to explicate with an equation. In different contexts the term may have special uses.

· In a section on frequently misused words in his book The Writer"s Art, James J. Kilpatrick quoted a letter from a correspondent, giving examples to illustrate the correct use of the word Parameter

· A parametric equaliser is an audio filter that allows the frequency of maximum cut or boost to be set by one control, and the size of the cut or boost by another.

Integrating factor

In mathematics, an integrating factor is a function that is chosen to facilitate the solving of a given equation involving differentials. It is commonly used to solve ordinary differential equations, but is also used within multivariable calculus, in this case often multiplying through by an integrating factor allows an inexact differential to be made into an exact differential (which can then be integrated to give a scalar field).

Consider an ordinary differential equation of the form

$$y' + P(x)y = Q(x) \qquad (1)$$

where y = y(x) is an unknown function of x, and P(x) and Q(x) are given functions.

Nominative determinism

Nominative determinism refers to the theory that a person"s name is given an influential role in reflecting key attributes of his job, profession, but real examples are more highly prized, the more obscure the better.

Weak solution

In mathematics, a weak solution to an ordinary or partial differential equation is a function for which the derivatives appearing in the equation may not all exist but which is nonetheless deemed to satisfy the equation in some precisely defined sense. There are many different definitions of weak solution, appropriate for different classes of equations. One of the most important is based on the notion of distributions.

Singular point

A singular point on a curve is one where it is not smooth, for example, at a cusp.

The precise definition of a singular point depends on the type of curve being studied.

Algebraic curves in R^2 are defined as the zero set $f^{-1}(0)$ for a polynomial function $f: R^2 \to R$. The singular point s are those points on the curve where both partial derivatives vanish,

$$f(x,y) = \frac{\partial f}{\partial x} = \frac{\partial f}{\partial y} = 0.$$

Chapter 2. FIRST-ORDER DIFFERENTIAL EQUATIONS

A parameterized curve in R² is defined as the image of a function g:R→R², g(t) = (g_1(t),g_2(t).)

Discontinuous

In mathematics, a continuous function is a function for which, intuitively, small changes in the input result in small changes in the output. Otherwise, a function is said to be discontinuous. A continuous function with a continuous inverse function is called bicontinuous.

Complementary error function

In mathematics, the error function is a special function (non-elementary) of sigmoid shape which occurs in probability, statistics, materials science, and partial differential equations. It is defined as:

$$\mathrm{erf}(x) = \frac{2}{\sqrt{\pi}} \int_0^x e^{-t^2} dt.$$

The Complementary error function, denoted erfc, is defined in terms of the error function:

$$\mathrm{erfc}(x) = 1 - \mathrm{erf}(x)$$
$$= \frac{2}{\sqrt{\pi}} \int_x^\infty e^{-t^2} dt.$$

The complex error function, denoted w(x), (also known as the Faddeeva function) is also defined in terms of the error function:

$$w(x) = e^{-x^2} \mathrm{erfc}(-ix).$$

Fig.2. Integrand exp(−z²) in the complex z-plane.

The error function is odd:

$$\mathrm{erf}(-z) = -\mathrm{erf}(z).$$

Also, for any complex number z:

$$\mathrm{erf}(\overline{z}) = \overline{\mathrm{erf}(z)}$$

where $\overline{z}$ is the complex conjugate of z.

The integrand $f = \exp(-z^2)$ and $f = \mathrm{erf}(z)$ are shown in the complex z-plane in figures 2 and 3. Level of Im(f) = 0 is shown with a thick green line.

Error function

In mathematics, the Error function is a special function (non-elementary) of sigmoid shape which occurs in probability, statistics, materials science, and partial differential equations. It is defined as:

$$\mathrm{erf}(x) = \frac{2}{\sqrt{\pi}} \int_0^x e^{-t^2} dt.$$

The complementary Error function, denoted erfc, is defined in terms of the Error function:

$$\mathrm{erfc}(x) = 1 - \mathrm{erf}(x)$$
$$= \frac{2}{\sqrt{\pi}} \int_x^\infty e^{-t^2} dt.$$

The complex Error function, denoted w(x), (also known as the Faddeeva function) is also defined in terms of the Error function:

$$w(x) = e^{-x^2}\operatorname{erfc}(-ix).$$

Fig.2. Integrand exp(−z²) in the complex z-plane.
The Error function is odd:
$$\operatorname{erf}(-z) = -\operatorname{erf}(z).$$
Also, for any complex number z:
$$\operatorname{erf}(\overline{z}) = \overline{\operatorname{erf}(z)}$$
where $\overline{z}$ is the complex conjugate of z.
The integrand $f = \exp(-z^2)$ and $f = \operatorname{erf}(z)$ are shown in the complex z-plane in figures 2 and 3. Level of Im(f) = 0 is shown with a thick green line.

Function	In mathematics, a function is a relation between a given set of elements (the domain) and another set of elements (the codomain), which associates each element in the domain with exactly one element in the codomain. The elements so related can be any kind of thing (words, objects, qualities) but are typically mathematical quantities, such as real numbers.
	There are many ways to represent or visualize functions: a function may be described by a formula, by a plot or graph, by an algorithm that computes it, by arrows between objects, or by a description of its properties.
Special functions	Special functions are particular mathematical functions which have more or less established names and notations due to their importance in mathematical analysis, functional analysis, physics, or other applications.
	There is no general formal definition, but the list of mathematical functions contains functions which are commonly accepted as special. In particular, elementary functions are also considered as Special functions
Defined	In mathematics, defined and undefined are used to explain whether or not expressions have meaningful, sensible, and unambiguous values. Whether an expression has a meaningful value depends on the context of the expression. For example the value of 4 − 5 is undefined if an positive integer result is required.
Input	Input is the term denoting either an entrance or changes which are inserted into a system and which activate/modify a process. It is an abstract concept, used in the modeling, system(s) design and system(s) expwdlfhaslfnasl"fnsfasnfloitation. It is usually connected with other terms, e.g., Input field, Input variable, Input parameter, Input value, Input signal, Input device and Input file.
Sine	The sine of an angle is the ratio of the length of the opposite side to the length of the hypotenuse. In our case $$\sin A = \frac{\text{opposite}}{\text{hypotenuse}} = \frac{a}{h}.$$ Note that this ratio does not depend on size of the particular right triangle chosen, as long as it contains the angle A, since all such triangles are similar.
	The co sine of an angle is the ratio of the length of the adjacent side to the length of the hypotenuse.

Chapter 2. FIRST-ORDER DIFFERENTIAL EQUATIONS

Sine integral

In mathematics, the trigonometric integrals are a family of integrals which involve trigonometric functions. A number of the basic trigonometric integrals are discussed at the list of integrals of trigonometric functions. Plot of Si(x) for $0 \le x \le 8\pi$. The different Sine integral definitions are:

$$\text{Si}(x) = \int_0^x \frac{\sin t}{t}\, dt$$

$$\text{si}(x) = -\int_x^\infty \frac{\sin t}{t}\, dt$$

Si(x) is the primitive of sinx / x which is zero for x = 0; si(x) is the primitive of sinx / x which is zero for $x = \infty$.

Series

In mathematics, given an infinite sequence of numbers $\{a_n\}$, a Series is informally the result of adding all those terms together: $a_1 + a_2 + a_3 + \cdots$. These can be written more compactly using the summation symbol Σ. An example is the famous Series from Zeno"s Dichotomy

$$\sum_{n=1}^\infty \frac{1}{2^n} = \frac{1}{2} + \frac{1}{4} + \frac{1}{8} + \cdots + \frac{1}{2^n} + \cdots$$

The terms of the Series are often produced according to a certain rule, such as by a formula, by an algorithm, by a sequence of measurements, or even by a random number generator.

Differential of a function

The differential is defined in modern treatments of differential calculus as follows. The Differential of a function $f(x)$ of a single real variable x is the function df of two independent real variables x and Δx given by

$$df(x, \Delta x) \overset{\text{def}}{=} f'(x)\, \Delta x.$$

One or both of the arguments may be suppressed, i.e., one may see df(x) or simply df. If y = f(x), the differential may also be written as dy.

Exact differential equation

In mathematics, an Exact differential equation or total differential equation is a certain kind of ordinary differential equation which is widely used in physics and engineering.

Given a simply connected and open subset D of R^2 and two functions I and J which are continuous on D then an implicit first-order ordinary differential equation of the form

$$I(x, y)\, dx + J(x, y)\, dy = 0,$$

is called Exact differential equation if there exists a continuously differentiable function F, called the potential function, so that

$$\frac{\partial F}{\partial x}(x, y) = I$$

and

$$\frac{\partial F}{\partial y}(x, y) = J.$$

Chapter 2. FIRST-ORDER DIFFERENTIAL EQUATIONS

The nomenclature of "Exact differential equation" refers to the exact derivative (or total derivative) of a function. For a function $F(x_0, x_1, ... x_{n-1}, x_n)$, the exact or total derivative with respect to x_0 is given by

$$\frac{dF}{dx_0} = \frac{\partial F}{\partial x_0} + \sum_{i=1}^{n} \frac{\partial F}{\partial x_i} \frac{dx_i}{dx_0}.$$

The function

$$F(x, y) := \frac{1}{2}(x^2 + y^2)$$

is a potential function for the differential equation

$$xx' + yy' = 0.$$

In physical applications the functions I and J are usually not only continuous but even continuously differentiable.

Binary function

In mathematics, a Binary function, is a function which takes two inputs.

Precisely stated, a function f is binary if there exists sets X,Y,Z such that

$$f: X \times Y \to Z$$

where $X \times Y$ is the Cartesian product of X and Y.

For example, if Z is the set of integers, N^+ is the set of natural numbers (except for zero), and Q is the set of rational numbers, then division is a Binary function from Z and N^+ to Q.

Set-theoretically, one may represent a Binary function as a subset of the Cartesian product X × Y × Z, where (x,y,z) belongs to the subset if and only if f(x,y) = z. Conversely, a subset R defines a Binary function if and only if, for any x in X and y in Y, there exists a unique z in Z such that (x,y,z) belongs to R. We then define f(x,y) to be this z.

Chain

A Chain is a series of connected links A Chain may consist of two or more links.

Homogeneous function

In mathematics, a Homogeneous function is a function with multiplicative scaling behaviour: if the argument is multiplied by a factor, then the result is multiplied by some power of this factor.

Suppose that $f : V \to W$ is a function between two vector spaces over a field F.

We say that f is homogeneous of degree k if

$$f(\alpha \mathbf{v}) = \alpha^k f(\mathbf{v})$$

for all nonzero $\alpha \in F$ and $\mathbf{v} \in V$.

Degree

In topology, the term degree is applied to continuous maps between manifolds of the same dimension. The degree of a map can be defined in terms of homology groups or, for smooth maps, in terms of preimages of regular values. It is a generalization of winding number.

Reduction

In mathematics, Reduction refers to the rewriting of an expression into a simpler form. For example, the process of rewriting a fraction into one with the smallest whole-number denominator possible (while keeping the numerator an integer) is called "reducing a fraction". Rewriting a radical (or "root") expression with the smallest possible whole number under the radical symbol is called "reducing a radical".

Chapter 2. FIRST-ORDER DIFFERENTIAL EQUATIONS

Separation of variables

In mathematics, Separation of variables is any of several methods for solving ordinary and partial differential equations, in which algebra allows one to rewrite an equation so that each of two variables occurs on a different side of the equation. Suppose a differential equation can be written in the form

$$\frac{d}{dx}f(x) = g(x)h(f(x)), \qquad (1)$$

which we can write more simply by letting y = f(x):

$$\frac{dy}{dx} = g(x)h(y).$$

As long as h(y) ≠ 0, we can rearrange terms to obtain:

$$\frac{dy}{h(y)} = g(x)dx,$$

so that the two variables x and y have been separated.

Some who dislike Leibniz"s notation may prefer to write this as

$$\frac{1}{h(y)}\frac{dy}{dx} = g(x),$$

but that fails to make it quite as obvious why this is called "Separation of variables".

Tangent

In geometry, the Tangent line (or simply the Tangent to a curve at a given point is the straight line that "just touches" the curve at that point (in the sense explained more precisely below.) As it passes through the point of tangency, the Tangent line is "going in the same direction" as the curve, and in this sense it is the best straight-line approximation to the curve at that point. The same definition applies to space curves and curves in n-dimensional Euclidean space.

Tangent line

In geometry, the tangent line (or simply the tangent) to a curve at a given point is the straight line that "just touches" the curve at that point (in the sense explained more precisely below.) As it passes through the point of tangency, the tangent line is "going in the same direction" as the curve, and in this sense it is the best straight-line approximation to the curve at that point. The same definition applies to space curves and curves in n-dimensional Euclidean space.

Linearization

Visually, the accompanying diagram shows the tangent line of f(x) at x. At f(x + h), where h is any small positive or negative value, f(x+h) is very nearly the value of the tangent line at the point (x + h,L(x + h)).

The final equation for the Linearization of a function at x = a is:

$$y = f(a) + f'(a)(x - a)$$

Chapter 2. FIRST-ORDER DIFFERENTIAL EQUATIONS

For x = a, f(a) = f(x).

Absolute error

The approximation error in some data is the discrepancy between an exact value and some approximation to it. An approximation error can occur because

· the measurement of the data is not precise (due to the instruments), or
· approximations are used instead of the real data (e.g., 3.14 instead of π).

In the mathematical field of numerical analysis, the numerical stability of an algorithm in numerical analysis indicates how the error is propagated by the algorithm.
One commonly distinguishes between the relative error and the Absolute error.

Percentage

In mathematics, a Percentage is a way of expressing a number as a fraction of 100 (per cent meaning "per hundred".) It is often denoted using the percent sign, "%". For example, 45% (read as "forty-five percent") is equal to 45 / 100, or 0.45.

Relative error

One commonly distinguishes between the relative error and the absolute error. The absolute error is the magnitude of the difference between the exact value and the approximation. The relative error is the absolute error divided by the magnitude of the exact value.

Runge-Kutta method

In mathematics, the Runge - Kutta method is a technique for the approximate numerical solution of a stochastic differential equation. It is a generalization of the Runge-Kutta method for ordinary differential equations to stochastic differential equations.

Consider the Itô diffusion X satisfying the following Itô stochastic differential equation

$$dX_t = a(X_t)\,dt + b(X_t)\,dW_t,$$

with initial condition $X_0 = x_0$, where W_t stands for the Wiener process, and suppose that we wish to solve this SDE on some interval of time [0, T].

Chapter 3. MODELING WITH FIRST-ORDER DIFFERENTIAL EQUATIONS

Differential equation — A Differential equation is a mathematical equation for an unknown function of one or several variables that relates the values of the function itself and its derivatives of various orders. Differential equations play a prominent role in engineering, physics, economics and other disciplines. Visualization of airflow into a duct modelled using the Navier-Stokes equations, a set of partial Differential equations.

Differential equations arise in many areas of science and technology: whenever a deterministic relationship involving some continuously changing quantities (modelled by functions) and their rates of change (expressed as derivatives) is known or postulated.

Mathematical model — Note: The term model has a different meaning in model theory, a branch of mathematical logic. An artifact which is used to illustrate a mathematical idea is also called a Mathematical model and this usage is the reverse of the sense explained below.

A Mathematical model uses mathematical language to describe a system.

Exponential growth — exponential growth (including exponential decay) occurs when the growth rate of a mathematical function is proportional to the function"s current value. In the case of a discrete domain of definition with equal intervals it is also called geometric growth or geometric decay (the function values form a geometric progression).

The exponential growth model is also known as the Malthusian growth model.

· Biology

· The number of microorganisms in a culture broth will grow exponentially until an essential nutrient is exhausted.

Linear — In a different usage to the above, a polynomial of degree 1 is said to be Linear, because the graph of a function of that form is a line.

Over the reals, a Linear equation is one of the form:
f(x) = m x + b
m is often called the slope or gradient; b the y-intercept, which gives the point of intersection between the graph of the function and the y-axis.

Note that this usage of the term Linear is not the same as the above, because Linear polynomials over the real numbers do not in general satisfy either additivity or homogeneity.

Ordinary differential equation — In mathematics, an ordinary differential equation (or ordinary differential equation) is a relation that contains functions of only one independent variable, and one or more of its derivatives with respect to that variable.

A simple example is Newton"s second law of motion, which leads to the differential equation

$$m\frac{d^2 x(t)}{dt^2} = F(x(t)),$$

for the motion of a particle of constant mass m. In general, the force F depends upon the position of the particle x(t) at time t, and thus the unknown function x(t) appears on both sides of the differential equation, as is indicated in the notation F(x(t).)

Decay constant — A quantity is said to be subject to exponential decay if it decreases at a rate proportional to its value. Symbolically, this can be expressed as the following differential equation, where N is the quantity and λ is a positive number called the Decay constant.

Chapter 3. MODELING WITH FIRST-ORDER DIFFERENTIAL EQUATIONS

$$\frac{dN}{dt} = -\lambda N.$$

The solution to this equation is:

$$N(t) = N_0 e^{-\lambda t}.$$

Here N(t) is the quantity at time t, and N_0 = N(0) is the initial quantity, i.e. the quantity at time t = 0.

Series	In mathematics, given an infinite sequence of numbers {a_n}, a Series is informally the result of adding all those terms together: $a_1 + a_2 + a_3 + \cdots$. These can be written more compactly using the summation symbol Σ. An example is the famous Series from Zeno"s Dichotomy $$\sum_{n=1}^{\infty} \frac{1}{2^n} = \frac{1}{2} + \frac{1}{4} + \frac{1}{8} + \cdots + \frac{1}{2^n} + \cdots$$ The terms of the Series are often produced according to a certain rule, such as by a formula, by an algorithm, by a sequence of measurements, or even by a random number generator.
Air	The Earth"s atmosphere is a layer of gases surrounding the planet Earth that is retained by Earth"s gravity. The atmosphere protects life on Earth by absorbing ultraviolet solar radiation, warming the surface through heat retention (greenhouse effect), and reducing temperature extremes between day and night. Dry Air contains roughly (by volume) 78.08% nitrogen, 20.95% oxygen, 0.93% argon, 0.038% carbon dioxide, and trace amounts of other gases.
State	In functional analysis, a State on a C*-algebra is a positive linear functional of norm 1. The set of States of a C*-algebra A, sometimes denoted by S(A), is always a convex set. The extremal points of S(A) are called pure States. If A has a multiplicative identity, S(A) is compact in the weak*-topology.
Continuous	In mathematics, a continuous function is a function for which, intuitively, small changes in the input result in small changes in the output. Otherwise, a function is said to be discontinuous. A continuous function with a continuous inverse function is called bicontinuous.
Compound interest	Compound interest arises when interest is added to the principal, so that from that moment on, the interest that has been added also itself earns interest. This addition of interest to the principal is called compounding (i.e. the interest is compounded). A loan, for example, may have its interest compounded every month: in this case, a loan with $100 initial principal and 1% interest per month would have a balance of $101 at the end of the first month, $102.01 at the end of the second month, and so on.
Tim	TIM (standing for "Task-oriented Information Modelling") designates a formal language engineered to allow the description of working situations by means of a task tree and to structure the information related of the tasks included in it. TIM models can provide the content for product-related documents (such as user-manuals, on-line helps and maintenance manuals) but also represent processes or express the functional specifications of hardware and software systems alongside with main features of their interfaces. TIM has been engineered by Tanguy Wettengel (PhD in Computational Linguistics, Higher Doctorate in Language Sciences) at the Centre de Recherches Sémiotiques (University of Limoges, France).

Chapter 3. MODELING WITH FIRST-ORDER DIFFERENTIAL EQUATIONS

Variable	A Variable is a symbol that stands for a value that may vary; the term usually occurs in opposition to constant, which is a symbol for a non-varying value, i.e. completely fixed or fixed in the context of use. The concepts of constants and variables are fundamental to all modern mathematics, science, engineering, and computer programming.
	Much of the basic theory for which we use variables today, such as school geometry and algebra, was developed thousands of years ago, but the use of symbolic formulae and variables is only several hundreds of years old.
Terminal velocity	In fluid dynamics an object is moving at its terminal velocity if its speed is constant due to the restraining force exerted by the air, water or other fluid through which it is moving.
	A free-falling object achieves its terminal velocity when the downward force of gravity (F_g) equals the upward force of drag (F_d). This causes the net force on the object to be zero, resulting in an acceleration of zero.
Plane	In mathematics, a plane is a flat surface. planes can arise as subspaces of some higher dimensional space, as with the walls of a room, or they may enjoy an independent existence in their own right, as in the setting of Euclidean geometry
Carrying capacity	The carrying capacity of a biological species in an environment is the population size of the species that the environment can sustain indefinitely, given the food, habitat, water and other necessities available in the environment. For the human population, more complex variables such as sanitation and medical care are sometimes considered as part of the necessary infrastructure.
	As population density increases, birth rate often increases and death rate typically decreases.
Logistic function	A Logistic function or logistic curve is the most common sigmoid curve. It models the "S-shaped" curve of growth of some set P, where P might be thought of as population. The initial stage of growth is approximately exponential; then, as saturation begins, the growth slows, and at maturity, growth stops.
Curve	In mathematics, a Curve consists of the points through which a continuously moving point passes. This notion captures the intuitive idea of a geometrical one-dimensional object, which furthermore is connected in the sense of having no discontinuities or gaps. Simple examples include the sine wave as the basic Curve underlying simple harmonic motion, and the parabola.
Maxima	In mathematics, maxima and minima, known collectively as extrema (singular: extremum), are the largest value (maximum) or smallest value (minimum), that a function takes in a point either within a given neighbourhood (local extremum) or on the function domain in its entirety (global extremum).
	Throughout, a point refers to an input (x), while a value refers to an output (y): one distinguishing between the maximum value and the point (or points) at which it occurs.
	A real-valued function f defined on the real line is said to have a local (or relative) maximum point at the point x^*, if there exists some $\varepsilon > 0$, such that $f(x^*) \geq f(x)$ when $|x - x^*| < \varepsilon$.
Action	In physics, action is an attribute of the development of a physical system. It is a functional which takes the trajectory (also called path or history) of the system as its argument and returns a real number as the result.
	It has units of energy × time (joule-seconds in SI units).

Chapter 3. MODELING WITH FIRST-ORDER DIFFERENTIAL EQUATIONS

Chapter 3. MODELING WITH FIRST-ORDER DIFFERENTIAL EQUATIONS

Shape

The Shape of an object located in some space is the part of that space occupied by the object, as determined by its external boundary - abstracting from other properties such as colour, content, and material composition, as well as from the object"s other spatial properties .

Mathematician and statistician David George Kendall defined Shape this way:

Shape is all the geometrical information that remains when location, scale and rotational effects are filtered out from an object.

Simple two-dimensional Shapes can be described by basic geometry such as points, line, curves, plane, and so on.

Surface

In mathematics, specifically in topology, a surface is a two-dimensional topological manifold. The most familiar examples are those that arise as the boundaries of solid objects in ordinary three-dimensional Euclidean space R^3 -- for example, the surface of a ball or bagel. On the other hand, there are surface s which cannot be embedded in three-dimensional Euclidean space without introducing singularities or intersecting itself -- these are the unorientable surface s.

Linear regression

In statistics, linear regression refers to any approach to modeling the relationship between one or more variables denoted y and one or more variables denoted X, such that the model depends linearly on the unknown parameters to be estimated from the data. Such a model is called a "linear model." Most commonly, linear regression refers to a model in which the conditional mean of y given the value of X is an affine function of X. Less commonly, linear regression could refer to a model in which the median, or some other quantile of the conditional distribution of y given X is expressed as a linear function of X. Like all forms of regression analysis, linear regression focuses on the conditional probability distribution of y given X, rather than on the joint probability distribution of y and X, which is the domain of multivariate analysis.

linear regression was the first type of regression analysis to be studied rigorously, and to be used extensively in practical applications.

Volterra integral equation

In mathematics, the Volterra integral equations are a special type of integral equations. They are divided into two groups referred to as the first and the second kind.

A linear Volterra equation of the first kind is

$$f(t) = \int_a^t K(t,s)\, x(s)\, ds.$$

A linear Volterra equation of the second kind is

$$x(t) = f(t) + \int_a^t K(t,s) x(s)\, ds.$$

In operator theory, and in Fredholm theory, the corresponding equations are called the Volterra operator.

Linear system

A linear system is a mathematical model of a system based on the use of a linear operator. linear systems typically exhibit features and properties that are much simpler than the general, nonlinear case. As a mathematical abstraction or idealization, linear systems find important applications in automatic control theory, signal processing, and telecommunications.

Chapter 2. FIRST-ORDER DIFFERENTIAL EQUATIONS

Chapter 3. MODELING WITH FIRST-ORDER DIFFERENTIAL EQUATIONS

Chapter 3. MODELING WITH FIRST-ORDER DIFFERENTIAL EQUATIONS

Nonlinear	In mathematics, a nonlinear system is a system which is not linear, that is, a system which does not satisfy the superposition principle, a nonlinear system is any problem where the variable(s) to be solved for cannot be written as a linear combination of independent components. A nonhomogeneous system, which is linear apart from the presence of a function of the independent variables, is nonlinear according to a strict definition, but such systems are usually studied alongside linear systems, because they can be transformed to a linear system of multiple variables.
Nonlinear system	In mathematics, a Nonlinear system is a system which is not linear, that is, a system which does not satisfy the superposition principle, a Nonlinear system is any problem where the variable(s) to be solved for cannot be written as a linear combination of independent components. A nonhomogeneous system, which is linear apart from the presence of a function of the independent variables, is nonlinear according to a strict definition, but such systems are usually studied alongside linear systems, because they can be transformed to a linear system of multiple variables.
Number	The use of zero as a number should be distinguished from its use as a placeholder numeral in place-value systems. Many ancient texts used zero. Babylonians and Egyptian texts used it.
Branch point	In the mathematical field of complex analysis, a Branch point of a multi-valued function (usually referred to as a "multifunction" in the context of complex analysis) is a point such that the function is discontinuous when going around an arbitrarily small circuit around this point (Ablowitz ' Fokas 2003, p. 46). Multi-valued functions are rigorously studied using Riemann surfaces, and the formal definition of Branch points employs this concept. Branch points fall into three broad categories: algebraic Branch points, transcendental Branch points, and logarithmic Branch points.
Second	The second (SI symbol: s), sometimes abbreviated sec., is the name of a unit of time, and is the International System of Units (SI) base unit of time. It may be measured using a clock. SI prefixes are frequently combined with the word second to denote subdivisions of the second e.g., the milli second (one thousandth of a second , the micro second (one millionth of a second , and the nano second (one billionth of a second)
Disease	A Disease or medical condition is an abnormal condition of an organism that impairs bodily functions, associated with specific symptoms and signs. It may be caused by external factors, such as infectious Disease, or it may be caused by internal dysfunctions, such as autoimmune Diseases. In human beings, "Disease" is often used more broadly to refer to any condition that causes pain, dysfunction, distress, social problems, and/or death to the person afflicted, or similar problems for those in contact with the person.
Cycloid	A Cycloid is the curve defined by the path of a point on the edge of circular wheel as the wheel rolls along a straight line. It is an example of a roulette, a curve generated by a curve rolling on another curve. The Cycloid is the solution to the brachistochrone problem (i.e. it is the curve of fastest descent under gravity) and the related tautochrone problem (i.e. the period of a ball rolling back and forth inside this curve does not depend on the ball"s starting position).

Chapter 3. MODELING WITH FIRST-ORDER DIFFERENTIAL EQUATIONS

Chapter 3. MODELING WITH FIRST-ORDER DIFFERENTIAL EQUATIONS

Tractrix — Tractrix is the curve along which a small object moves, under the influence of friction, when pulled on a horizontal plane by a piece of thread and a puller that moves at a right angle to the initial line between the object and the puller at an infinitesimal speed. It is therefore a curve of pursuit. It was first introduced by Claude Perrault in 1670, and later studied by Sir Isaac Newton and Christian Huygens (1692.)

Orthogonal — In mathematics, two vectors are orthogonal if they are perpendicular, i.e., they form a right angle. The word comes from the Greek á½€ρθÏŒς , meaning "straight", and γωνÍ α (gonia), meaning "angle". For example, a subway and the street above, although they do not physically intersect, are orthogonal if they cross at a right angle.

Orthogonal trajectories — In mathematics, orthogonal trajectories are a family of curves in the plane that intersect a given family of curves at right angles. The problem is classical, but is now understood by means of complex analysis; see for example harmonic conjugate.

For a family of level curves described by g(x,y) = C, where C is a constant, the orthogonal trajectories may be found as the level curves of a new function f(x,y) by solving the partial differential equation

$$\nabla f \cdot \nabla g = 0$$

for f(x,y).

Chapter 4. HIGHER-ORDER DIFFERENTIAL EQUATIONS

Linear

In a different usage to the above, a polynomial of degree 1 is said to be Linear, because the graph of a function of that form is a line.

Over the reals, a Linear equation is one of the form:

$f(x) = m x + b$

m is often called the slope or gradient; b the y-intercept, which gives the point of intersection between the graph of the function and the y-axis.

Note that this usage of the term Linear is not the same as the above, because Linear polynomials over the real numbers do not in general satisfy either additivity or homogeneity.

Differential equation

A Differential equation is a mathematical equation for an unknown function of one or several variables that relates the values of the function itself and its derivatives of various orders. Differential equations play a prominent role in engineering, physics, economics and other disciplines. Visualization of airflow into a duct modelled using the Navier-Stokes equations, a set of partial Differential equations.

Differential equations arise in many areas of science and technology: whenever a deterministic relationship involving some continuously changing quantities (modelled by functions) and their rates of change (expressed as derivatives) is known or postulated.

Order

Order refers to a large number of concepts in mathematics, especially in algebra, arithmetic, analysis, combinatorics, fractals, graphs, and mathematical theories.

· Order of computation, the computational complexity of an algorithm
· Canonical Order, the Order of elements that obeys a certain set of rules or specifications
· Z-Order, the Order of windows on computer screens

· First-Order hold in signal processing
· Modulation Order, the number of different symbols that can be sent using a given modulation
· The polynomial Order of a filter transfer function

· Money Order
· Order (business), an instruction from a customer to buy
· Order (exchange), customer"s instruction to a stock broker
· Order of degrees in the Elliott wave principle

· Court Order, made by a judge; a restraining Order, for example, is a type of injunction
· Executive Order, issued by the executive branch of government
· General Order, a published directive from a commander
· Law and Order (politics)
· Order, or military command
· Social Order, referring to the conduct of society
· Standing Order, a general Order of indefinite duration, and similar ongoing rules in a parliament
· World Order, including the concept of a world government

· Architectonic Orders: see classical Order
· Public Order, a concept in urban planning

· Order (decoration), medal or award

Chapter 4. HIGHER-ORDER DIFFERENTIAL EQUATIONS

Chapter 4. HIGHER-ORDER DIFFERENTIAL EQUATIONS

- Chivalric Order, established since the 14th century
- Fraternal Order
- Holy Orders, the rite or sacrament in which clergy are ordained
- Military Order, established in the crusades
- Monastic Order, established since circa 300 AD
- Religious Order
- Order (organization), an organization of people united by a common fraternal bond or social aim
- Tariqa or Sufi Order
- Cardassian military unit in the fictional Star Trek universe
- Order of the Mishnah, the name given to a sub-division of the Mishnah, a major religious text
- Order of the Mass is the set of texts of the Roman Catholic Church Latin Rite Mass that are generally invariable.
- The Order (group), an underground American neo-Nazi organization active in 1983 and 1984.

Ordinary differential equation

In mathematics, an ordinary differential equation (or ordinary differential equation) is a relation that contains functions of only one independent variable, and one or more of its derivatives with respect to that variable.

A simple example is Newton"s second law of motion, which leads to the differential equation

$$m\frac{d^2x(t)}{dt^2} = F(x(t)),$$

for the motion of a particle of constant mass m. In general, the force F depends upon the position of the particle x(t) at time t, and thus the unknown function x(t) appears on both sides of the differential equation, as is indicated in the notation F(x(t).)

Initial condition

In mathematics, in the field of differential equations, an initial value problem is an ordinary differential equation together with specified value, called the initial condition, of the unknown function at a given point in the domain of the solution. In physics or other sciences, modeling a system frequently amounts to solving an initial value problem; in this context, the differential equation is an evolution equation specifying how, given initial conditions, the system will evolve with time.

An initial value problem is a differential equation

$$y'(t) = f(t, y(t)) \quad \text{with} \quad f : \mathbb{R} \times \mathbb{R} \to \mathbb{R}$$

together with a point in the domain of f

$$(t_0, y_0) \in \mathbb{R} \times \mathbb{R},$$

called the initial condition.

Initial value problem

In mathematics, in the field of differential equations, an initial value problem is an ordinary differential equation together with specified value, called the initial condition, of the unknown function at a given point in the domain of the solution. In physics or other sciences, modeling a system frequently amounts to solving an initial value problem; in this context, the differential equation is an evolution equation specifying how, given initial conditions, the system will evolve with time.

An initial value problem is a differential equation

$$y'(t) = f(t, y(t)) \quad \text{with} \quad f : \mathbb{R} \times \mathbb{R} \to \mathbb{R}$$

together with a point in the domain of f

Chapter 4. HIGHER-ORDER DIFFERENTIAL EQUATIONS

Chapter 4. HIGHER-ORDER DIFFERENTIAL EQUATIONS

$$(t_0, y_0) \in \mathbb{R} \times \mathbb{R},$$

called the initial condition.

Interval

In mathematics, a (real) interval is a set of real numbers with the property that any number that lies between two numbers in the set is also included in the set. For example, the set of all numbers x satisfying $0 \leq x \leq 1$ is an interval which contains 0 and 1, as well as all numbers between them. Other examples of intervals are the set of all real numbers $\mathbb{R}$, the set of all negative real numbers, and the empty set.

Uniqueness

In mathematics and logic, the phrase "there is one and only one" is used to indicate that exactly one object with a certain property exists. In mathematical logic, this sort of quantification is known as Uniqueness quantification or unique existential quantification.

Uniqueness quantification is often denoted with the symbols "∃!" or ∃$_{=1}$".

Derivative

In calculus (a branch of mathematics) the Derivative is a measure of how a function changes as its input changes. Loosely speaking, a Derivative can be thought of as how much a quantity is changing at a given point; for example, the Derivative of the position of a vehicle with respect to time is the instantaneous velocity at which the vehicle is traveling. Conversely, the integral of the velocity over time is the change in the vehicle"s position.

Theorem

In mathematics, a Theorem is a statement proved on the basis of previously accepted or established statements such as axioms. In formal mathematical logic, the concept of a Theorem may be taken to mean a formula that can be derived according to the derivation rules of a fixed formal system. The statements of a theory as expressed in a formal language are called its elementary Theorem s and are said to be true.

Boundary conditions

In mathematics, in the field of differential equations, a boundary value problem is a differential equation together with a set of additional restraints, called the boundary conditions. A solution to a boundary value problem is a solution to the differential equation which also satisfies the boundary conditions.

Boundary value problems arise in several branches of physics as any physical differential equation will have them.

Mathematical model

Note: The term model has a different meaning in model theory, a branch of mathematical logic. An artifact which is used to illustrate a mathematical idea is also called a Mathematical model and this usage is the reverse of the sense explained below.

A Mathematical model uses mathematical language to describe a system.

Homogeneous differential equation

A Homogeneous differential equation has several distinct meanings.

One meaning is that a first-order ordinary differential equation is homogeneous if it has the form

$$\frac{dy}{dx} = F(y/x).$$

To solve such equations, one makes the change of variables u = y/x, which will transform such an equation into separable one.

Another meaning is a linear Homogeneous differential equation, which is a differential equation of the form

$$Ly = 0$$

Chapter 4. HIGHER-ORDER DIFFERENTIAL EQUATIONS

Chapter 4. HIGHER-ORDER DIFFERENTIAL EQUATIONS

where the differential operator L is a linear operator, and y is the unknown function.

Linear differential equation

In mathematics, a Linear differential equation is of the form

$$Ly = f$$

where the differential operator L is a linear operator, y is the unknown function (such as a function of time y(t)), and the right hand side f is a given function of the same nature as y (called the source term). For a function dependent on time we may write the equation more expressively as

$$Ly(t) = f(t)$$

and, even more precisely by bracketing

$$L[y(t)] = f(t)$$

The linear operator L may be considered to be of the form

$$L_n(y) \equiv \frac{d^n y}{dt^n} + A_1(t)\frac{d^{n-1}y}{dt^{n-1}} + \cdots + A_{n-1}(t)\frac{dy}{dt} + A_n(t)y$$

The linearity condition on L rules out operations such as taking the square of the derivative of y; but permits, for example, taking the second derivative of y. It is convenient to rewrite this equation in an operator form

$$L_n(y) \equiv \left[D^n + A_1(t)D^{n-1} + \cdots + A_{n-1}(t)D + A_n(t)\right]y$$

where D is the differential operator d/dt (i.e. Dy = y'', D^2y = y'',...

Differential operator

In mathematics, a Differential operator is an operator defined as a function of the differentiation operator. It is helpful, as a matter of notation first, to consider differentiation as an abstract operation, accepting a function and returning another (in the style of a higher-order function in computer science).

There are certainly reasons not to restrict to linear operators; for instance the Schwarzian derivative is a well-known non-linear operator.

Linear operator

In mathematics, a linear map (, linear function) is a function between two vector spaces that preserves the operations of vector addition and scalar multiplication. The expression "linear operator" is commonly used for linear maps from a vector space to itself (endomorphisms). In advanced mathematics, the definition of linear function coincides with the definition of linear map.

Superposition principle

In physics and systems theory, the Superposition principle also known as superposition property, states that, for all linear systems,

> The net response at a given place and time caused by two or more stimuli is the sum of the responses which would have been caused by each stimulus individually.

So that if input A produces response X and input B produces response Y then input (A + B) produces response (X + Y.) Mathematically, for all linear systems F(x) = y, where x is some sort of stimulus (input) and y is some sort of response (output), the superposition (i.e., sum) of stimuli yields a superposition of the respective responses:

$$F(x_1 + x_2 + \cdots) = F(x_1) + F(x_2) + \cdots$$

Chapter 4. HIGHER-ORDER DIFFERENTIAL EQUATIONS

Chapter 4. HIGHER-ORDER DIFFERENTIAL EQUATIONS

Function

In mathematics, a function is a relation between a given set of elements (the domain) and another set of elements (the codomain), which associates each element in the domain with exactly one element in the codomain. The elements so related can be any kind of thing (words, objects, qualities) but are typically mathematical quantities, such as real numbers. There are many ways to represent or visualize functions: a function may be described by a formula, by a plot or graph, by an algorithm that computes it, by arrows between objects, or by a description of its properties.

Wronskian

In mathematics, the Wronskian is a function especially important in the study of differential equations, where it can be used to determine whether a set of solutions is linearly independent.

For n real- or complex-valued functions $f_1, ..., f_n$, which are n − 1 times differentiable on an interval I, the Wronskian $W(f_1, ..., f_n)$ as a function on I is defined by

$$W(f_1, \ldots, f_n)(x) = \begin{vmatrix} f_1(x) & f_2(x) & \cdots & f_n(x) \\ f_1'(x) & f_2'(x) & \cdots & f_n'(x) \\ \vdots & \vdots & \ddots & \vdots \\ f_1^{(n-1)}(x) & f_2^{(n-1)}(x) & \cdots & f_n^{(n-1)}(x) \end{vmatrix}, \quad x \in I.$$

That is, it is the determinant of the matrix constructed by placing the functions in the first row, the first derivative of each function in the second row, and so on through the (n - 1)st derivative, thus forming a square matrix sometimes called a fundamental matrix.

Gematria

Gematria or gimatria is a system of assigning numerical value to a word or phrase, in the belief that words or phrases with identical numerical values bear some relation to each other, the calendar year, or the like. The word "Gematria" is generally held to derive from Greek geÅ metriÄ , "geometry", which was used a translation of gÄ"maá¹riyÄ , though some scholars believe it to derive from Greek grammateia, rather; it"s possible that both words had an influence on the formation of the Hebrew word. It has been extant in English since the 17th century from translations of works by Giovanni Pico della Mirandola.

Nominative determinism

Nominative determinism refers to the theory that a person"s name is given an influential role in reflecting key attributes of his job, profession, but real examples are more highly prized, the more obscure the better.

Weak solution

In mathematics, a weak solution to an ordinary or partial differential equation is a function for which the derivatives appearing in the equation may not all exist but which is nonetheless deemed to satisfy the equation in some precisely defined sense. There are many different definitions of weak solution, appropriate for different classes of equations. One of the most important is based on the notion of distributions.

Linear system

A linear system is a mathematical model of a system based on the use of a linear operator. linear systems typically exhibit features and properties that are much simpler than the general, nonlinear case. As a mathematical abstraction or idealization, linear systems find important applications in automatic control theory, signal processing, and telecommunications.

Chapter 4. HIGHER-ORDER DIFFERENTIAL EQUATIONS

Chapter 4. HIGHER-ORDER DIFFERENTIAL EQUATIONS

State
In functional analysis, a State on a C*-algebra is a positive linear functional of norm 1. The set of States of a C*-algebra A, sometimes denoted by S(A), is always a convex set. The extremal points of S(A) are called pure States. If A has a multiplicative identity, S(A) is compact in the weak*-topology.

State variable
A state variable is one of the set of variables that describe the "state" of a dynamical system. Intuitively, the state of a system describes enough about the system to determine its future behaviour.

In simple mechanical systems, position coordinates and their derivatives are typical state variables; knowing these, it is possible to determine the future positions of objects in the system (if the higher order derivatives are zero).

Reduction
In mathematics, Reduction refers to the rewriting of an expression into a simpler form. For example, the process of rewriting a fraction into one with the smallest whole-number denominator possible (while keeping the numerator an integer) is called "reducing a fraction". Rewriting a radical (or "root") expression with the smallest possible whole number under the radical symbol is called "reducing a radical".

Reduction of order
Reduction of order is a technique in mathematics for solving second-order ordinary differential equations. It is employed when one solution $y_1(x)$ is known and a second linearly independent solution $y_2(x)$ is desired.

Consider the general second-order constant coefficient ODE

$$ay''(x) + by'(x) + cy(x) = 0,$$

where a,b,c are real non-zero coefficients.

Riccati
Jacopo Francesco Riccati (28 May 1676 - 15 April 1754) was an Italian mathematician, born in Venice. He is now remembered for the Riccati equation. He died in Treviso in 1754.

He received his early education at the Jesuit school for the nobility in Brescia.

Uniform Acceleration
uniform Acceleration is a type of motion in which the velocity of an object changes equal amounts in equal time periods. An example of an object having uniform Acceleration would be a ball rolling down a ramp. The object picks up velocity as it goes down the ramp with equal changes in time.

Formula

In mathematics, a Formula is a key to solve an equation with variables. For example, the problem of determining the volume of a sphere is one that requires a significant amount of integral calculus to solve. However, having done this once, mathematicians can produce a Formula to describe the volume in terms of some other parameter (the radius for example).

Linear equation
A Linear equation is an algebraic equation in which each term is either a constant or the product of a constant and (the first power of) a single variable.

Linear equations can have one or more variables. Linear equations occur with great regularity in applied mathematics.

Rational root
In algebra, the Rational root theorem (or "Rational root test") states a constraint on rational solutions (or roots) of the polynomial equation

$$a_n x^n + a_{n-1} x^{n-1} + \cdots + a_0 = 0$$

Chapter 4. HIGHER-ORDER DIFFERENTIAL EQUATIONS

with integer coefficients.

If a_0 and a_n are nonzero, then each rational solution x, when written as a fraction x = p/q in lowest terms (i.e., the greatest common divisor of p and q is 1) satisfies

· p is an integer factor of the constant term a_0, and
· q is an integer factor of the leading coefficient a_n.

Thus, a list of possible Rational roots of the equation can be derived using the formula
$$x = \pm \frac{p}{q}.$$

For example, every rational solution of the equation
$$3x^3 - 5x^2 + 5x - 2 = 0$$
must be among the numbers symbolically indicated by
$$\pm \frac{1,2}{1,3},$$
which gives the list of possible answers:
$$1, -1, 2, -2, \frac{1}{3}, -\frac{1}{3}, \frac{2}{3}, -\frac{2}{3}.$$

These root candidates can be tested, for example using the Horner scheme.

Synthetic division

Synthetic division is a method of performing polynomial long division, with less writing and fewer calculations. It is mostly taught for division by binomials of the form

x − a,

but the method generalizes to division by any monic polynomial. This method can be used instead of long division on integers by considering 10 = x and only substituting 10 back in at the end.

Method of undetermined coefficients

In mathematics, the method of undetermined coefficients is an approach to finding a particular solution to certain inhomogeneous ordinary differential equations and recurrence relations. It is closely related to the annihilator method, but instead of using a particular kind of differential operator (the annihilator) in order to find the best possible form of the particular solution, a "guess" is made as to the appropriate form, which is then tested by differentiating the resulting equation. In this sense, the method of undetermined coefficients is less formal but more intuitive than the annihilator method.

Variation of parameters

In mathematics, variation of parameters also known as variation of constants, is a general method to solve inhomogeneous linear ordinary differential equations. It was developed by the Italian-French mathematician Joseph Louis Lagrange with noteworthy help from the American mathematician and physicist Noah LaMoyne.

For first-order inhomogeneous linear differential equations it"s usually possible to find solutions via integrating factors or undetermined coefficients with considerably less effort, although those methods are rather heuristics that involve guessing and don"t work for all inhomogenous linear differential equations.

Parameter

In mathematics, statistics, and the mathematical sciences, a Parameter is a quantity that defines certain characteristics of systems or functions that serves to relate functions and variables using a common variable (often t) when such a relationship would be difficult to explicate with an equation. In different contexts the term may have special uses.

Chapter 4. HIGHER-ORDER DIFFERENTIAL EQUATIONS

Chapter 4. HIGHER-ORDER DIFFERENTIAL EQUATIONS

· In a section on frequently misused words in his book The Writer''s Art, James J. Kilpatrick quoted a letter from a correspondent, giving examples to illustrate the correct use of the word Parameter

· A parametric equaliser is an audio filter that allows the frequency of maximum cut or boost to be set by one control, and the size of the cut or boost by another.

Elementary

In computational complexity theory, the complexity class ELEMENTARY is the union of the classes in the exponential hierarchy.

$$\begin{aligned}\text{ELEMENTARY} &= \text{EXP} \cup \text{2EXP} \cup \text{3EXP} \cup \cdots \\ &= \text{DTIME}(2^n) \cup \text{DTIME}(2^{2^n}) \cup \text{DTIME}(2^{2^{2^n}}) \cup \cdots\end{aligned}$$

The name was coined by Laszlo Kalmar, in the context of recursive functions and undecidability; most problems in it are far from ELEMENTARY. Some natural recursive problems lie outside ELEMENTARY, and are thus NONELEMENTARY. Most notably, there are primitive recursive problems which are not in ELEMENTARY. We know

LOWER-ELEMENTARY $\subsetneq$ EXPTIME $\subsetneq$ ELEMENTARY $\subsetneq$ PR

Whereas ELEMENTARY contains bounded applications of exponentiation (for example, $O(2^{2^n})$), PR allows more general hyper operators (for example, tetration) which are not contained in ELEMENTARY.

The definitions of ELEMENTARY recursive functions are the same as for primitive recursive functions, except that primitive recursion is replaced by bounded summation and bounded product.

Elementary function

In mathematics, an Elementary function is a function built from a finite number of exponentials, logarithms, constants, one variable, and nth roots through composition and combinations using the four elementary operations (+ - × ÷). By allowing these functions .

The roots of equations are the functions implicitly defined as solving a polynomial equation with constant coefficients.

Laplace transform

In mathematics, the Laplace transform is a widely used integral transform. It has many important applications in mathematics, physics, optics, electrical engineering, control engineering, signal processing, and probability theory.

The Laplace transform is related to the Fourier transform, but whereas the Fourier transform resolves a function or signal into its modes of vibration, the Laplace transform resolves a function into its moments.

Projectile motion

projectile motion is one of the traditional branches of classical mechanics, with applications to ballistics. A projectile is any body that is given an initial velocity and then follows a path determined by the effect of the gravitational acceleration and by air resistance. For example, a thrown football, an object dropped from an airplane, or a bullet shot from a gun are all examples of projectiles.

Taylor series

In mathematics, the Taylor series is a representation of a function as an infinite sum of terms calculated from the values of its derivatives at a single point. It may be regarded as the limit of the Taylor polynomials. If the series is centered at zero, the series is also called a Maclaurin series.

Chapter 4. HIGHER-ORDER DIFFERENTIAL EQUATIONS

Taylor's theorem	In calculus, Taylor''s theorem gives a sequence of approximations of a differentiable function around a given point by polynomials (the Taylor polynomials of that function) whose coefficients depend only on the derivatives of the function at that point. The theorem also gives precise estimates on the size of the error in the approximation. The theorem is named after the mathematician Brook Taylor, who stated it in 1712, though the result was first discovered 41 years earlier in 1671 by James Gregory.
Curvature	In mathematics, Curvature refers to any of a number of loosely related concepts in different areas of geometry. Intuitively, Curvature is the amount by which a geometric object deviates from being flat, or straight in the case of a line, but this is defined in different ways depending on the context. There is a key distinction between extrinsic Curvature, which is defined for objects embedded in another space (usually a Euclidean space) in a way that relates to the radius of Curvature of circles that touch the object, and intrinsic Curvature, which is defined at each point in a differential manifold.

Chapter 5. MODELING WITH HIGHER-ORDER DIFFERENTIAL EQUATIONS

Differential equation	A Differential equation is a mathematical equation for an unknown function of one or several variables that relates the values of the function itself and its derivatives of various orders. Differential equations play a prominent role in engineering, physics, economics and other disciplines. Visualization of airflow into a duct modelled using the Navier-Stokes equations, a set of partial Differential equations.
	Differential equations arise in many areas of science and technology: whenever a deterministic relationship involving some continuously changing quantities (modelled by functions) and their rates of change (expressed as derivatives) is known or postulated.
Eigenfunction	In mathematics, an Eigenfunction of a linear operator, A, defined on some function space is any non-zero function f in that space that returns from the operator exactly as is, except for a multiplicative scaling factor. More precisely, one has $$\mathcal{A}f = \lambda f$$ for some scalar, λ, the corresponding eigenvalue. The solution of the differential eigenvalue problem also depends on any boundary conditions required of f.
Eigenvalue	In mathematics, eigenvalue, eigenvector, and eigenspace are related concepts in the field of linear algebra. Linear algebra studies linear transformations, which are represented by matrices acting on vectors. eigenvalues, eigenvectors and eigenspaces are properties of a matrix.
Mathematical model	Note: The term model has a different meaning in model theory, a branch of mathematical logic. An artifact which is used to illustrate a mathematical idea is also called a Mathematical model and this usage is the reverse of the sense explained below.
	A Mathematical model uses mathematical language to describe a system.
Acceleration	acceleration is the rate of change of velocity. At any point on a trajectory, the magnitude of the acceleration is given by the rate of change of velocity in both magnitude and direction at that point. The true acceleration at time t is found in the limit as time interval Δt → 0. Components of acceleration for a planar curved motion.
Equilibrium	Kreps and Robert Wilson · Symmetric Equilibrium, in game theory, an Equilibrium where all players use the same strategy · Trembling hand perfect Equilibrium assumes that the players, through a "slip of the hand" or tremble, may choose unintended strategies
	· Proper Equilibrium due to Roger B. Myerson, where costly trembles are made with smaller probabilities

Chapter 5. MODELING WITH HIGHER-ORDER DIFFERENTIAL EQUATIONS

Chapter 5. MODELING WITH HIGHER-ORDER DIFFERENTIAL EQUATIONS

- Equilibrium (band), folk metal band from Germany
- Equilibrium, a 2000 album by Crowbar
- Equilibrium , a software services and development company
- Equilibrium (film), a 2002 science-fiction film
- "Equilibrium" (seaQuest 2032 episode), episode of seaQuest 2032
- "Equilibrium" (DS9 episode), an episode of Star Trek: Deep Space Nine
- Equilibrium moisture content, the moisture content at which the wood is neither gaining nor losing moisture
- Equilibrium point, node in mathematics
- Equilibrium (puzzle), a 3D interlocking type puzzle made of 6 half circle pieces
- Reflective Equilibrium, the state of balance or coherence among a set of beliefs arrived at by a process of deliberative mutual adjustment
- Social Equilibrium, a system in which there is a dynamic working balance among its interdependent parts
- Equilibration is the means of cognitive development in Jean Piaget"s cognitive developmental theory. .

Input

Input is the term denoting either an entrance or changes which are inserted into a system and which activate/modify a process. It is an abstract concept, used in the modeling, system(s) design and system(s) expwdlfhaslfnasl"fnsfasnfloitation. It is usually connected with other terms, e.g., Input field, Input variable, Input parameter, Input value, Input signal, Input device and Input file.

Linear

In a different usage to the above, a polynomial of degree 1 is said to be Linear, because the graph of a function of that form is a line.

Over the reals, a Linear equation is one of the form:

f(x) = m x + b

m is often called the slope or gradient; b the y-intercept, which gives the point of intersection between the graph of the function and the y-axis.

Note that this usage of the term Linear is not the same as the above, because Linear polynomials over the real numbers do not in general satisfy either additivity or homogeneity.

Spring

In geometry, a Spring is a surface of revolution in the shape of a helix with thickness, generated by revolving a circle about the path of a helix. The torus is a special case of the Spring obtained when the helix is crushed to a circle.

A Spring wrapped around the z-axis can be defined parametrically by:

$$x(u,v) = (R + r\cos v)\cos u,$$
$$y(u,v) = (R + r\cos v)\sin u,$$
$$z(u,v) = r\sin v + \frac{P \cdot u}{\pi},$$

Chapter 5. MODELING WITH HIGHER-ORDER DIFFERENTIAL EQUATIONS

where

$u \in [0,\ 2n\pi]\ (n \in \mathbb{R})$,
$v \in [0,\ 2\pi]$,
R is the distance from the center of the tube to the center of the helix,
r is the radius of the tube,
P is the speed of the movement along the z axis (in a right-handed Cartesian coordinate system, positive values create right-handed springs, whereas negative values create left-handed springs),
n is the number of rounds in circle.

Spring constant

The most commonly encountered form of Hooke"s law is probably the spring equation, which relates the force exerted by a spring to the distance it is stretched by a spring constant, k, measured in force per length.

$$F = -kx$$

The negative sign indicates that the force exerted by the spring is in direct opposition to the direction of displacement. It is called a "restoring force", as it tends to restore the system to equilibrium.

Gravitation

Gravitation is a natural phenomenon by which objects with mass attract one another. In everyday life, Gravitation is most commonly thought of as the agency which lends weight to objects with mass. Gravitation causes dispersed matter to coalesce, thus accounting for the existence of the Earth, the Sun, and most of the macroscopic objects in the universe.

Linear model

In statistics, the term linear model is used in different ways according to the context. The most common occurrence is in connection with regression models and the term is often taken as synonymous with linear regression model. However the term is also used in time series analysis with a different meaning.

Ordinary differential equation

In mathematics, an ordinary differential equation (or ordinary differential equation) is a relation that contains functions of only one independent variable, and one or more of its derivatives with respect to that variable.

A simple example is Newton"s second law of motion, which leads to the differential equation

$$m\frac{d^2 x(t)}{dt^2} = F(x(t)),$$

for the motion of a particle of constant mass m. In general, the force F depends upon the position of the particle x(t) at time t, and thus the unknown function x(t) appears on both sides of the differential equation, as is indicated in the notation F(x(t).)

Second

The second (SI symbol: s), sometimes abbreviated sec., is the name of a unit of time, and is the International System of Units (SI) base unit of time. It may be measured using a clock.

SI prefixes are frequently combined with the word second to denote subdivisions of the second e.g., the milli second (one thousandth of a second , the micro second (one millionth of a second , and the nano second (one billionth of a second)

Chapter 5. MODELING WITH HIGHER-ORDER DIFFERENTIAL EQUATIONS

Chapter 5. MODELING WITH HIGHER-ORDER DIFFERENTIAL EQUATIONS

Hyperbolic cosine	In mathematics, the hyperbolic functions are analogs of the ordinary trigonometric, functions. The basic hyperbolic functions are the hyperbolic sine "sinh", and the hyperbolic cosine "cosh", from which are derived the hyperbolic tangent "tanh", etc., in analogy to the derived trigonometric functions. The inverse hyperbolic functions are the area hyperbolic sine "arsinh" (also called "asinh", or sometimes by the misnomer of "arcsinh") and so on.
Equation	An Equation is a mathematical statement, in symbols, that two things are exactly the same (or equivalent). Equations are written with an equal sign, as in $$2 + 3 = 5.$$ $$9 - 2 = 7.$$ The Equations above are examples of an equality: a proposition which states that two constants are equal. Equalities may be true or false.
Frequency	Frequency is the number of occurrences of a repeating event per unit time. It is also referred to as temporal Frequency. The period is the duration of one cycle in a repeating event, so the period is the reciprocal of the Frequency.
Simple harmonic motion	simple harmonic motion is the motion of a simple harmonic oscillator, a motion that is neither driven nor damped. The motion is periodic - as it repeats itself at standard intervals in a specific manner - and sinusoidal, with constant amplitude; the acceleration of a body executing simple harmonic motion is directly proportional to the displacement of the body from the equilibrium position and is always directed towards the equilibrium position. The motion is characterized by its amplitude (which is always positive), its period, the time for a single oscillation, its frequency, the reciprocal of the period (i.e. the number of cycles per unit time), and its phase, which determines the starting point on the sine wave.
Phase angle	The notation $A \angle \theta$, for a vector with magnitude (or amplitude) A and Phase angle θ, is called angle notation. In the context of periodic phenomena, such as a wave, Phase angle is synonymous with phase.
Variable	A Variable is a symbol that stands for a value that may vary; the term usually occurs in opposition to constant, which is a symbol for a non-varying value, i.e. completely fixed or fixed in the context of use. The concepts of constants and variables are fundamental to all modern mathematics, science, engineering, and computer programming. Much of the basic theory for which we use variables today, such as school geometry and algebra, was developed thousands of years ago, but the use of symbolic formulae and variables is only several hundreds of years old.
Damping	In physics, damping is any effect that tends to reduce the amplitude of oscillations in an oscillatory system, particularly the harmonic oscillator. In mechanics, friction is one such damping effect. For many purposes the frictional force F_f can be modeled as being proportional to the velocity v of the object: $F_f = -cv$, where c is the viscous damping coefficient, given in units of newton-seconds per meter.
Resonance	In physics, Resonance is the tendency of a system to oscillate at larger amplitude at some frequencies than at others. These are known as the system"s resonant frequencies (or Resonance frequencies). At these frequencies, even small periodic driving forces can produce large amplitude vibrations, because the system stores vibrational energy.

Chapter 5. MODELING WITH HIGHER-ORDER DIFFERENTIAL EQUATIONS

Chapter 5. MODELING WITH HIGHER-ORDER DIFFERENTIAL EQUATIONS

Series

In mathematics, given an infinite sequence of numbers {a_n}, a Series is informally the result of adding all those terms together: $a_1 + a_2 + a_3 + \cdots$. These can be written more compactly using the summation symbol Σ. An example is the famous Series from Zeno"s Dichotomy

$$\sum_{n=1}^{\infty} \frac{1}{2^n} = \frac{1}{2} + \frac{1}{4} + \frac{1}{8} + \cdots + \frac{1}{2^n} + \cdots$$

The terms of the Series are often produced according to a certain rule, such as by a formula, by an algorithm, by a sequence of measurements, or even by a random number generator.

Air

The Earth"s atmosphere is a layer of gases surrounding the planet Earth that is retained by Earth"s gravity. The atmosphere protects life on Earth by absorbing ultraviolet solar radiation, warming the surface through heat retention (greenhouse effect), and reducing temperature extremes between day and night. Dry Air contains roughly (by volume) 78.08% nitrogen, 20.95% oxygen, 0.93% argon, 0.038% carbon dioxide, and trace amounts of other gases.

Reactance

Reactance is a circuit element"s opposition to an alternating current, caused by the build up of electric or magnetic fields in the element due to the current. Both fields act to produce counter EMF that is proportional to either the rate of change (time derivative), or accumulation (time integral) of the current. In vector analysis, Reactance is the imaginary part of electrical impedance, used to compute amplitude and phase changes of sinusoidal alternating current going through the circuit element.

State

In functional analysis, a State on a C*-algebra is a positive linear functional of norm 1. The set of States of a C*-algebra A, sometimes denoted by S(A), is always a convex set. The extremal points of S(A) are called pure States. If A has a multiplicative identity, S(A) is compact in the weak*-topology.

Curve

In mathematics, a Curve consists of the points through which a continuously moving point passes. This notion captures the intuitive idea of a geometrical one-dimensional object, which furthermore is connected in the sense of having no discontinuities or gaps. Simple examples include the sine wave as the basic Curve underlying simple harmonic motion, and the parabola.

Curvature

In mathematics, Curvature refers to any of a number of loosely related concepts in different areas of geometry. Intuitively, Curvature is the amount by which a geometric object deviates from being flat, or straight in the case of a line, but this is defined in different ways depending on the context. There is a key distinction between extrinsic Curvature, which is defined for objects embedded in another space (usually a Euclidean space) in a way that relates to the radius of Curvature of circles that touch the object, and intrinsic Curvature, which is defined at each point in a differential manifold.

Flexural rigidity

Flexural rigidity is defined as the force couple required to bend a rigid structure to a unit curvature.

In a beam or rod, Flexural rigidity varies along the length as a function of x shown in the following equation:

$$EI\frac{dy}{dx} = \int_0^x M(x)dx + C_1$$

where E is the modulus of elasticity, I is the 2nd moment of inertia, y is the transverse displacement of the beam at x, and M(x) is the bending moment at x.

Chapter 5. MODELING WITH HIGHER-ORDER DIFFERENTIAL EQUATIONS

Chapter 5. MODELING WITH HIGHER-ORDER DIFFERENTIAL EQUATIONS

Flexural rigidity has SI units of $Pa \cdot m^4$ (which also equals $N \cdot m^2$).

Symmetry	symmetry generally conveys two primary meanings. The first is an imprecise sense of harmonious or aesthetically-pleasing proportionality and balance; such that it reflects beauty or perfection. The second meaning is a precise and well-defined concept of balance or "patterned self-similarity" that can be demonstrated or proved according to the rules of a formal system: by geometry, through physics or otherwise.
Boundary conditions	In mathematics, in the field of differential equations, a boundary value problem is a differential equation together with a set of additional restraints, called the boundary conditions. A solution to a boundary value problem is a solution to the differential equation which also satisfies the boundary conditions.
Boundary value problems arise in several branches of physics as any physical differential equation will have them.
Critical load	In the study of air pollution, a Critical load is defined as "A quantitative estimate of an exposure to one or more pollutants below which significant harmful effects on specified sensitive elements of the environment do not occur according to present knowledge". (Nilsson and Grennfelt 1988)
Critical loads and the similar concept of critical levels have been used extensively within the 1979 UN-ECE Convention on Long-Range Transboundary Air Pollution. As an example the 1999 Gothenburg protocol to the LRTAP convention takes into account acidification (of surface waters and soils), eutrophication of soils and ground-level ozone and the emissions of sulfur dioxide, ammonia, nitrogen oxide and non-methane volatile organic compounds (NMVOC).
Vertical	A pair of angles is said to be Vertical or opposite if the angles share the same vertex and are bounded by the same pair of lines but are opposite to each other. Such angles are congruent and thus have equal measure. If two line segments, EF and GH, intersect at the point P, they form four angles, EPG, GPF, FPH, and HPE.
Critical speed	In Solid mechanics, in the field of rotordynamics, the Critical speed is the theoretical angular velocity which excites the natural frequency of a rotating object, such as a shaft, propeller or gear. As the speed of rotation approaches the objects"s natural frequency, the object begins to resonate which dramatically increases systemic vibration. The resulting resonance occurs regardless of orientation.
Periodic boundary conditions	In mathematical models and computer simulations, Periodic boundary conditions are a set of boundary conditions that are often used to simulate a large system by modelling a small part that is far from its edge. Periodic boundary conditions resemble the topologies of some video games; a unit cell or simulation box of a geometry suitable for perfect three-dimensional tiling is defined, and when an object passes through one face of the unit cell, it reappears on the opposite face with the same velocity. The simulation is of an infinite perfect tiling of the system.
Sphere	A Sphere is a perfectly round geometrical object in three-dimensional space, such as the shape of a round ball. Like a circle in two dimensions, a perfect Sphere is completely symmetrical around its center, with all points on the surface lying the same distance r from the center point. This distance r is known as the radius of the Sphere

Chapter 5. MODELING WITH HIGHER-ORDER DIFFERENTIAL EQUATIONS

Chapter 5. MODELING WITH HIGHER-ORDER DIFFERENTIAL EQUATIONS

Nonlinear

In mathematics, a nonlinear system is a system which is not linear, that is, a system which does not satisfy the superposition principle, a nonlinear system is any problem where the variable(s) to be solved for cannot be written as a linear combination of independent components. A nonhomogeneous system, which is linear apart from the presence of a function of the independent variables, is nonlinear according to a strict definition, but such systems are usually studied alongside linear systems, because they can be transformed to a linear system of multiple variables.

Linearization

Visually, the accompanying diagram shows the tangent line of f(x) at x. At f(x + h), where h is any small positive or negative value, f(x+h) is very nearly the value of the tangent line at the point (x + h, L(x + h)).

The final equation for the Linearization of a function at x = a is:

$$y = f(a) + f'(a)(x - a)$$

For x = a, f(a) = f(x).

Simple pendulum

The period of swing of a simple gravity pendulum depends on its length, the acceleration of gravity, and to a small extent on the maximum angle that the pendulum swings away from vertical, θ_0, called the amplitude. It is independent of the mass of the bob. If the amplitude is limited to small swings, the period T of a Simple pendulum, the time taken for a complete cycle, is:

$$T \approx 2\pi\sqrt{\frac{L}{g}} \qquad \theta_0 \ll 1 \qquad (1)$$

where L is the length of the pendulum and g is the local acceleration of gravity.

Catenary

In mathematics, a commutative ring R is catenary if for any pair of prime ideals

p, q,

any two strictly increasing chains

p=p$_0$ ⊂p$_1$ ... ⊂p$_n$= q of prime ideals

are contained in maximal strictly increasing chains from p to q of the same (finite) length. In other words, there is a well-defined function from pairs of prime ideals to natural numbers, attaching to p and q the length of any such maximal chain.

Shape

The Shape of an object located in some space is the part of that space occupied by the object, as determined by its external boundary - abstracting from other properties such as colour, content, and material composition, as well as from the object"s other spatial properties .

Mathematician and statistician David George Kendall defined Shape this way:

Shape is all the geometrical information that remains when location, scale and rotational effects are filtered out from an object.

Simple two-dimensional Shapes can be described by basic geometry such as points, line, curves, plane, and so on.

Suspension

In topology, the Suspension SX of a topological space X is the quotient space:

$$SX = (X \times I)/\{(x_1, 0) \sim (x_2, 0) \text{ and } (x_1, 1) \sim (x_2, 1) \text{ for all } x_1, x_2 \in X\}$$

Suspension of a circle. The original space is in blue, and the collapsed end points are in green.

Chapter 5. MODELING WITH HIGHER-ORDER DIFFERENTIAL EQUATIONS

Chapter 5. MODELING WITH HIGHER-ORDER DIFFERENTIAL EQUATIONS

of the product of X with the unit interval I = [0, 1]. Intuitively, we make X into a cylinder and collapse both ends to two points.

Gematria Gematria or gimatria is a system of assigning numerical value to a word or phrase, in the belief that words or phrases with identical numerical values bear some relation to each other, the calendar year, or the like. The word "Gematria" is generally held to derive from Greek geÅ metriÄ , "geometry", which was used a translation of gÄ"maá¹riyÄ , though some scholars believe it to derive from Greek grammateia, rather; it"s possible that both words had an influence on the formation of the Hebrew word. It has been extant in English since the 17th century from translations of works by Giovanni Pico della Mirandola.

Chain A Chain is a series of connected links A Chain may consist of two or more links.

Escape velocity In physics, Escape velocity is the speed where the kinetic energy of an object is equal to the magnitude of its gravitational potential energy, as calculated by the equation,

$$U_g = \frac{-G m_1 m_2}{r}$$

It is commonly described as the speed needed to "break free" from a gravitational field (without any additional impulse) and is theoretical, totally neglecting atmospheric friction. The term Escape velocity can be considered a misnomer because it is actually a speed rather than a velocity, i.e. it specifies how fast the object must move but the direction of movement is irrelevant, unless "downward." In more technical terms, Escape velocity is a scalar (and not a vector). Escape velocity gives a minimum delta-v budget for rockets when no benefit can be obtained from the speeds of other bodies for a particular mission; but it neglects losses such as air drag and gravity drag.

Chapter 5. MODELING WITH HIGHER-ORDER DIFFERENTIAL EQUATIONS

Chapter 6. SERIES SOLUTIONS OF LINEAR EQUATIONS

Interval

In mathematics, a (real) interval is a set of real numbers with the property that any number that lies between two numbers in the set is also included in the set. For example, the set of all numbers x satisfying $0 \leq x \leq 1$ is an interval which contains 0 and 1, as well as all numbers between them. Other examples of intervals are the set of all real numbers $\mathbb{R}$, the set of all negative real numbers, and the empty set.

Power series

In mathematics, a Power series (in one variable) is an infinite series of the form

$$f(x) = \sum_{n=0}^{\infty} a_n (x-c)^n = a_0 + a_1(x-c)^1 + a_2(x-c)^2 + a_3(x-c)^3 + \cdots$$

where a_n represents the coefficient of the nth term, c is a constant, and x varies around c (for this reason one sometimes speaks of the series as being centered at c

In many situations c is equal to zero, for instance when considering a Maclaurin series.

Radius of convergence

In mathematics, the Radius of convergence of a power series is a non-negative quantity, either a real number or ∞, that represents a domain (within the radius) in which the series will converge. Within the Radius of convergence a power series converges absolutely and uniformly on compacta as well. If the series converges, it is the Taylor series of the analytic function to which it converges inside its Radius of convergence

Ratio

A Ratio is an expression that compares quantities relative to each other. The most common examples involve two quantities, but any number of quantities can be compared. Ratio s are represented mathematically by separating each quantity with a colon - for example, the Ratio 2:3, which is read as the Ratio "two to three".

Ratio test

In mathematics, the ratio test is a test (or "criterion") for the convergence of a series

$$\sum_{n=0}^{\infty} a_n$$

whose terms are non-zero real or complex numbers. The test was first published by Jean le Rond d"Alembert and is sometimes known as d"Alembert"s ratio test The test makes use of the number

()

in the cases where this limit exists.

Series

In mathematics, given an infinite sequence of numbers {a_n}, a Series is informally the result of adding all those terms together: $a_1 + a_2 + a_3 + \cdots$. These can be written more compactly using the summation symbol Σ. An example is the famous Series from Zeno"s Dichotomy

$$\sum_{n=1}^{\infty} \frac{1}{2^n} = \frac{1}{2} + \frac{1}{4} + \frac{1}{8} + \cdots + \frac{1}{2^n} + \cdots$$

The terms of the Series are often produced according to a certain rule, such as by a formula, by an algorithm, by a sequence of measurements, or even by a random number generator.

Chapter 6. SERIES SOLUTIONS OF LINEAR EQUATIONS

Chapter 6. SERIES SOLUTIONS OF LINEAR EQUATIONS

Differential equation

A Differential equation is a mathematical equation for an unknown function of one or several variables that relates the values of the function itself and its derivatives of various orders. Differential equations play a prominent role in engineering, physics, economics and other disciplines. Visualization of airflow into a duct modelled using the Navier-Stokes equations, a set of partial Differential equations.

Differential equations arise in many areas of science and technology: whenever a deterministic relationship involving some continuously changing quantities (modelled by functions) and their rates of change (expressed as derivatives) is known or postulated.

Linear

In a different usage to the above, a polynomial of degree 1 is said to be Linear, because the graph of a function of that form is a line.

Over the reals, a Linear equation is one of the form:

f(x) = m x + b

m is often called the slope or gradient; b the y-intercept, which gives the point of intersection between the graph of the function and the y-axis.

Note that this usage of the term Linear is not the same as the above, because Linear polynomials over the real numbers do not in general satisfy either additivity or homogeneity.

Integration

Integration is an important concept in mathematics which, together with differentiation, forms one of the main operations in calculus. Given a function f of a real variable x and an interval [a, b] of the real line, the definite integral

$$\int_a^b f(x)\,dx,$$

is defined informally to be the net signed area of the region in the xy-plane bounded by the graph of f, the x-axis, and the vertical lines x = a and x = b.

The term integral may also refer to the notion of antiderivative, a function F whose derivative is the given function f.

Summation

Summation is the addition of a set of numbers; the result is their sum or total. An interim or present total of a Summation process is termed the running total. The "numbers" to be summed may be natural numbers, complex numbers, matrices, or still more complicated objects.

Gematria

Gematria or gimatria is a system of assigning numerical value to a word or phrase, in the belief that words or phrases with identical numerical values bear some relation to each other, the calendar year, or the like. The word "Gematria" is generally held to derive from Greek geÅ metriÄ , "geometry", which was used a translation of gÄ"maá†riyÄ , though some scholars believe it to derive from Greek grammateia, rather; it"s possible that both words had an influence on the formation of the Hebrew word. It has been extant in English since the 17th century from translations of works by Giovanni Pico della Mirandola.

Riccati

Jacopo Francesco Riccati (28 May 1676 - 15 April 1754) was an Italian mathematician, born in Venice. He is now remembered for the Riccati equation. He died in Treviso in 1754.

He received his early education at the Jesuit school for the nobility in Brescia.

Singular point

A singular point on a curve is one where it is not smooth, for example, at a cusp.

The precise definition of a singular point depends on the type of curve being studied.

Chapter 6. SERIES SOLUTIONS OF LINEAR EQUATIONS

Chapter 6. SERIES SOLUTIONS OF LINEAR EQUATIONS

Algebraic curves in R^2 are defined as the zero set $f^{-1}(0)$ for a polynomial function $f:R^2\to R$. The singular point s are those points on the curve where both partial derivatives vanish,

$$f(x,y) = \frac{\partial f}{\partial x} = \frac{\partial f}{\partial y} = 0.$$

A parameterized curve in R^2 is defined as the image of a function $g:R\to R^2$, $g(t) = (g_1(t), g_2(t).)$

Ordinary differential equation	In mathematics, an ordinary differential equation (or ordinary differential equation) is a relation that contains functions of only one independent variable, and one or more of its derivatives with respect to that variable. A simple example is Newton"s second law of motion, which leads to the differential equation $$m\frac{d^2 x(t)}{dt^2} = F(x(t)),$$ for the motion of a particle of constant mass m. In general, the force F depends upon the position of the particle x(t) at time t, and thus the unknown function x(t) appears on both sides of the differential equation, as is indicated in the notation F(x(t).)
Recurrence relation	In mathematics, a Recurrence relation is an equation that defines a sequence recursively: each term of the sequence is defined as a function of the preceding terms. A difference equation is a specific type of Recurrence relation An example of a Recurrence relation is the logistic map: $$x_{n+1} = r x_n (1 - x_n)$$ Some simply defined Recurrence relation s can have very complex (chaotic) behaviours and are sometimes studied by physicists and mathematicians in a field of mathematics known as nonlinear analysis.
Curve	In mathematics, a Curve consists of the points through which a continuously moving point passes. This notion captures the intuitive idea of a geometrical one-dimensional object, which furthermore is connected in the sense of having no discontinuities or gaps. Simple examples include the sine wave as the basic Curve underlying simple harmonic motion, and the parabola.
Irregular singular point	In mathematics, in the theory of ordinary differential equations in the complex plane $\mathbb{C}$, the points of $\mathbb{C}$ are classified into ordinary points, at which the equation"s coefficients are analytic functions, and singular points, at which some coefficient has a singularity. Then amongst singular points, an important distinction is made between a regular singular point, where the growth of solutions is bounded (in any small sector) by an algebraic function, and an irregular singular point, where the full solution set requires functions with higher growth rates. This distinction occurs, for example, between the hypergeometric equation, with three regular singular points, and the Bessel equation which is in a sense a limiting case, but where the analytic properties are substantially different.

Chapter 6. SERIES SOLUTIONS OF LINEAR EQUATIONS

Chapter 6. SERIES SOLUTIONS OF LINEAR EQUATIONS

Regular singular point

In mathematics, in the theory of ordinary differential equations in the complex plane $\mathbb{C}$, the points of $\mathbb{C}$ are classified into ordinary points, at which the equation"s coefficients are analytic functions, and singular points, at which some coefficient has a singularity. Then amongst singular points, an important distinction is made between a Regular singular point where the growth of solutions is bounded (in any small sector) by an algebraic function, and an ir Regular singular point , where the full solution set requires functions with higher growth rates. This distinction occurs, for example, between the hypergeometric equation, with three Regular singular point s, and the Bessel equation which is in a sense a limiting case, but where the analytic properties are substantially different.

Linear differential equation

In mathematics, a Linear differential equation is of the form

$$Ly = f$$

where the differential operator L is a linear operator, y is the unknown function (such as a function of time y(t)), and the right hand side f is a given function of the same nature as y (called the source term). For a function dependent on time we may write the equation more expressively as

$$Ly(t) = f(t)$$

and, even more precisely by bracketing

$$L[y(t)] = f(t)$$

The linear operator L may be considered to be of the form

$$L_n(y) \equiv \frac{d^n y}{dt^n} + A_1(t)\frac{d^{n-1}y}{dt^{n-1}} + \cdots + A_{n-1}(t)\frac{dy}{dt} + A_n(t)y$$

The linearity condition on L rules out operations such as taking the square of the derivative of y; but permits, for example, taking the second derivative of y. It is convenient to rewrite this equation in an operator form

$$L_n(y) \equiv \left[D^n + A_1(t)D^{n-1} + \cdots + A_{n-1}(t)D + A_n(t) \right] y$$

where D is the differential operator d/dt (i.e. Dy = y" , D^2y = y",...

Theorem

In mathematics, a Theorem is a statement proved on the basis of previously accepted or established statements such as axioms. In formal mathematical logic, the concept of a Theorem may be taken to mean a formula that can be derived according to the derivation rules of a fixed formal system. The statements of a theory as expressed in a formal language are called its elementary Theorem s and are said to be true.

Root

In vascular plants, the Root is the organ of a plant that typically lies below the surface of the soil. This is not always the case, however, since a Root can also be aerial (growing above the ground) or aerating (growing up above the ground or especially above water.) Furthermore, a stem normally occurring below ground is not exceptional either

Singularity

In mathematics, a singularity is in general a point at which a given mathematical object is not defined such as differentiability. See singularity theory for general discussion of the geometric theory, which only covers some aspects. For example, the function

$$f(x) = \frac{1}{x}$$

Chapter 6. SERIES SOLUTIONS OF LINEAR EQUATIONS

Chapter 6. SERIES SOLUTIONS OF LINEAR EQUATIONS

	on the real line has a singularity at $x = 0$, where it seems to "explode" to $\pm\infty$ and is not defined.
Nominative determinism	Nominative determinism refers to the theory that a person"s name is given an influential role in reflecting key attributes of his job, profession, but real examples are more highly prized, the more obscure the better.
Weak solution	In mathematics, a weak solution to an ordinary or partial differential equation is a function for which the derivatives appearing in the equation may not all exist but which is nonetheless deemed to satisfy the equation in some precisely defined sense. There are many different definitions of weak solution, appropriate for different classes of equations. One of the most important is based on the notion of distributions.
Order	Order refers to a large number of concepts in mathematics, especially in algebra, arithmetic, analysis, combinatorics, fractals, graphs, and mathematical theories. · Order of computation, the computational complexity of an algorithm · Canonical Order, the Order of elements that obeys a certain set of rules or specifications · Z-Order, the Order of windows on computer screens · First-Order hold in signal processing · Modulation Order, the number of different symbols that can be sent using a given modulation · The polynomial Order of a filter transfer function · Money Order · Order (business), an instruction from a customer to buy · Order (exchange), customer"s instruction to a stock broker · Order of degrees in the Elliott wave principle · Court Order, made by a judge; a restraining Order, for example, is a type of injunction · Executive Order, issued by the executive branch of government · General Order, a published directive from a commander · Law and Order (politics) · Order, or military command · Social Order, referring to the conduct of society · Standing Order, a general Order of indefinite duration, and similar ongoing rules in a parliament · World Order, including the concept of a world government · Architectonic Orders: see classical Order · Public Order, a concept in urban planning · Order (decoration), medal or award

Chapter 6. SERIES SOLUTIONS OF LINEAR EQUATIONS

Chapter 6. SERIES SOLUTIONS OF LINEAR EQUATIONS

- Chivalric Order, established since the 14th century
- Fraternal Order
- Holy Orders, the rite or sacrament in which clergy are ordained
- Military Order, established in the crusades
- Monastic Order, established since circa 300 AD
- Religious Order
- Order (organization), an organization of people united by a common fraternal bond or social aim
- Tariqa or Sufi Order
- Cardassian military unit in the fictional Star Trek universe
- Order of the Mishnah, the name given to a sub-division of the Mishnah, a major religious text
- Order of the Mass is the set of texts of the Roman Catholic Church Latin Rite Mass that are generally invariable.
- The Order (group), an underground American neo-Nazi organization active in 1983 and 1984.

Bessel functions

In mathematics, Bessel functions, first defined by the mathematician Daniel Bernoulli and generalized by Friedrich Bessel, are canonical solutions y(x) of Bessel"s differential equation:

$$x^2 \frac{d^2 y}{dx^2} + x \frac{dy}{dx} + (x^2 - \alpha^2) y = 0$$

for an arbitrary real or complex number α (the order of the Bessel function). The most common and important special case is where α is an integer n.

Although α and −α produce the same differential equation, it is conventional to define different Bessel functions for these two orders (e.g., so that the Bessel functions are mostly smooth functions of α).

Gamma function

In mathematics, the Gamma function is an extension of the factorial function to real and complex numbers. For a complex number z with positive real part the Gamma function is defined by

$$\Gamma(z) = \int_0^\infty t^{z-1} e^{-t}\, dt$$

This definition can be extended by analytic continuation to the rest of the complex plane, except the non-positive integers.
If n is a positive integer, then
Γ = (n − 1)!
showing the connection to the factorial function.

Spring

In geometry, a Spring is a surface of revolution in the shape of a helix with thickness, generated by revolving a circle about the path of a helix. The torus is a special case of the Spring obtained when the helix is crushed to a circle.

A Spring wrapped around the z-axis can be defined parametrically by:

$$\begin{aligned} x(u,v) &= (R + r \cos v) \cos u, \\ y(u,v) &= (R + r \cos v) \sin u, \\ z(u,v) &= r \sin v + \frac{P \cdot u}{\pi}, \end{aligned}$$

Chapter 6. SERIES SOLUTIONS OF LINEAR EQUATIONS

where

$u \in [0, \ 2n\pi] \ (n \in \mathbb{R})$,
$v \in [0, \ 2\pi]$,
R is the distance from the center of the tube to the center of the helix,
r is the radius of the tube,
P is the speed of the movement along the z axis (in a right-handed Cartesian coordinate system, positive values create right-handed springs, whereas negative values create left-handed springs),
n is the number of rounds in circle.

Graph	In mathematics, a graph is an abstract representation of a set of objects where some pairs of the objects are connected by links. The interconnected objects are represented by mathematical abstractions called vertices, and the links that connect some pairs of vertices are called edges. Typically, a graph is depicted in diagrammatic form as a set of dots for the vertices, joined by lines or curves for the edges.
Second	The second (SI symbol: s), sometimes abbreviated sec., is the name of a unit of time, and is the International System of Units (SI) base unit of time. It may be measured using a clock. SI prefixes are frequently combined with the word second to denote subdivisions of the second e.g., the milli second (one thousandth of a second , the micro second (one millionth of a second , and the nano second (one billionth of a second)
Mathematical model	Note: The term model has a different meaning in model theory, a branch of mathematical logic. An artifact which is used to illustrate a mathematical idea is also called a Mathematical model and this usage is the reverse of the sense explained below. A Mathematical model uses mathematical language to describe a system.
Uniform Acceleration	uniform Acceleration is a type of motion in which the velocity of an object changes equal amounts in equal time periods. An example of an object having uniform Acceleration would be a ball rolling down a ramp. The object picks up velocity as it goes down the ramp with equal changes in time.
Legendre functions	Note: People , Legendre functions are solutions to Legendre"s differential equation: $$\frac{d}{dx}\left[(1-x^2)\frac{d}{dx}P_n(x)\right] + n(n+1)P_n(x) = 0.$$ They are named after Adrien-Marie Legendre. This ordinary differential equation is frequently encountered in physics and other technical fields.
Mathieu functions	In mathematics, the Mathieu functions are certain special functions useful for treating a variety of problems in applied mathematics, including

Chapter 6. SERIES SOLUTIONS OF LINEAR EQUATIONS

Chapter 6. SERIES SOLUTIONS OF LINEAR EQUATIONS

- vibrating elliptical drumheads,
- quadrupoles mass filters and quadrupole ion traps for mass spectrometry
- the phenomenon of parametric resonance in forced oscillators,
- exact plane wave solutions in general relativity.
- the Stark effect for a rotating electric dipole.

They were introduced by Émile Léonard Mathieu in 1868 in the context of the first problem.

The canonical form for Mathieu"s differential equation is

$$\frac{d^2 u}{dx^2} + [a_u - 2q_u \cos(2x)]u = 0.$$

Closely related is Mathieu"s modified differential equation

$$\frac{d^2 y}{du^2} - [a - 2q \cosh(2u)]y = 0$$

which follows on substitution u = ix.

Special functions

Special functions are particular mathematical functions which have more or less established names and notations due to their importance in mathematical analysis, functional analysis, physics, or other applications.

There is no general formal definition, but the list of mathematical functions contains functions which are commonly accepted as special. In particular, elementary functions are also considered as Special functions

Formula

In mathematics, a Formula is a key to solve an equation with variables. For example, the problem of determining the volume of a sphere is one that requires a significant amount of integral calculus to solve. However, having done this once, mathematicians can produce a Formula to describe the volume in terms of some other parameter (the radius for example).

Chapter 7. THE LAPLACE TRANSFORM

Sawtooth	The sawtooth wave (or saw wave) is a kind of non-sinusoidal waveform. It is named a sawtooth based on its resemblance to the teeth on the blade of a saw.
	The convention is that a sawtooth wave ramps upward and then sharply drops.
Function	In mathematics, a function is a relation between a given set of elements (the domain) and another set of elements (the codomain), which associates each element in the domain with exactly one element in the codomain. The elements so related can be any kind of thing (words, objects, qualities) but are typically mathematical quantities, such as real numbers.
	There are many ways to represent or visualize functions: a function may be described by a formula, by a plot or graph, by an algorithm that computes it, by arrows between objects, or by a description of its properties.
Integral transform	In mathematics, an Integral transform is any transform T of the following form: $$Tf(u) = \int_{t_1}^{t_2} K(t,u)\, f(t)\, dt.$$
	The input of this transform is a function f, and the output is another function Tf. An Integral transform is a particular kind of mathematical operator.
	There are numerous useful Integral transforms.
Null space	In linear algebra, the kernel (also nullspace) of a matrix A is the set of all vectors x for which Ax = 0. The null space of a matrix with n columns is a linear subspace of n-dimensional Euclidean space.
	The nullspace (or kernel) of the matrix A is exactly the same thing as the nullspace (or kernel) of the linear mapping defined by the matrix-vector multiplication $\mathbf{x} \mapsto \mathbf{Ax}$, that is, the set of vectors that map to the zero vector.
	In linear algebra, the null space (also nullspace) of a matrix A is the set of all vectors x for which Ax = 0. The null space of a matrix with n columns is a linear subspace of n-dimensional Euclidean space.
Laplace transform	In mathematics, the Laplace transform is a widely used integral transform. It has many important applications in mathematics, physics, optics, electrical engineering, control engineering, signal processing, and probability theory.
	The Laplace transform is related to the Fourier transform, but whereas the Fourier transform resolves a function or signal into its modes of vibration, the Laplace transform resolves a function into its moments.
Definition	A Definition is a formal passage describing the meaning of a term (a word or phrase). The term to be defined is the definiendum. A term may have many subtly different senses or meanings.
Improper integral	In calculus, an Improper integral is the limit of a definite integral as an endpoint of the interval of integration approaches either a specified real number or ∞ or −∞ or, in some cases, as both endpoints approach limits.
	Specifically, an Improper integral is a limit of the form $$\lim_{b \to \infty} \int_a^b f(x)\, dx, \qquad \lim_{a \to -\infty} \int_a^b f(x)\, dx,$$ or of the form $$\lim_{c \to b^-} \int_a^c f(x)\, dx, \qquad \lim_{c \to a^+} \int_c^b f(x)\, dx,$$

Chapter 7. THE LAPLACE TRANSFORM

Chapter 7. THE LAPLACE TRANSFORM

in which one takes a limit in one or the other (or sometimes both) endpoints (Apostol 1967, §10.23). Improper integrals may also occur at an interior point of the domain of integration, or at multiple such points.

Property

Property is any physical or intangible entity that is owned by a person or jointly by a group of persons. Depending on the nature of the Property an owner of Property has the right to consume, sell, rent, mortgage, transfer, exchange or destroy his or her Property and/or to exclude others from doing these things. Important widely-recognized types of Property include real Property (land), personal Property (physical possessions belonging to an person), private Property Property owned by legal persons or business entities), public Property (state owned or publicly owned and available possessions) and intellectual Property (exclusive rights over artistic creations, inventions, etc.), although the latter is not always as widely recognized or enforced.

Linear

In a different usage to the above, a polynomial of degree 1 is said to be Linear, because the graph of a function of that form is a line.

Over the reals, a Linear equation is one of the form:

$f(x) = m x + b$

m is often called the slope or gradient; b the y-intercept, which gives the point of intersection between the graph of the function and the y-axis.

Note that this usage of the term Linear is not the same as the above, because Linear polynomials over the real numbers do not in general satisfy either additivity or homogeneity.

Order

Order refers to a large number of concepts in mathematics, especially in algebra, arithmetic, analysis, combinatorics, fractals, graphs, and mathematical theories.

· Order of computation, the computational complexity of an algorithm
· Canonical Order, the Order of elements that obeys a certain set of rules or specifications
· Z-Order, the Order of windows on computer screens

· First-Order hold in signal processing
· Modulation Order, the number of different symbols that can be sent using a given modulation
· The polynomial Order of a filter transfer function

· Money Order
· Order (business), an instruction from a customer to buy
· Order (exchange), customer''s instruction to a stock broker
· Order of degrees in the Elliott wave principle

· Court Order, made by a judge; a restraining Order, for example, is a type of injunction
· Executive Order, issued by the executive branch of government
· General Order, a published directive from a commander
· Law and Order (politics)
· Order, or military command
· Social Order, referring to the conduct of society
· Standing Order, a general Order of indefinite duration, and similar ongoing rules in a parliament
· World Order, including the concept of a world government

Chapter 7. THE LAPLACE TRANSFORM

Chapter 7. THE LAPLACE TRANSFORM

- Architectonic Orders: see classical Order
- Public Order, a concept in urban planning
- Order (decoration), medal or award
- Chivalric Order, established since the 14th century
- Fraternal Order
- Holy Orders, the rite or sacrament in which clergy are ordained
- Military Order, established in the crusades
- Monastic Order, established since circa 300 AD
- Religious Order
- Order (organization), an organization of people united by a common fraternal bond or social aim
- Tariqa or Sufi Order
- Cardassian military unit in the fictional Star Trek universe
- Order of the Mishnah, the name given to a sub-division of the Mishnah, a major religious text
- Order of the Mass is the set of texts of the Roman Catholic Church Latin Rite Mass that are generally invariable.
- The Order (group), an underground American neo-Nazi organization active in 1983 and 1984.

Sufficient

In propositional logic, a set of Boolean operators is called sufficient if it permits the realisation of any possible truth table.
Example truth table (Xor):
Using a complete Boolean algebra which does not include XOR (such as the well-known AND OR NOT set), this function can be realised as follows:
(a or b) and not (a and b.)
However, other complete Boolean algebras are possible, such as NAND or NOR (either gate can form a complete Boolean algebra by itself - the proof is detailed on their pages.)

Gamma function

In mathematics, the Gamma function is an extension of the factorial function to real and complex numbers. For a complex number z with positive real part the Gamma function is defined by

$$\Gamma(z) = \int_0^\infty t^{z-1} e^{-t}\, dt$$

This definition can be extended by analytic continuation to the rest of the complex plane, except the non-positive integers.
If n is a positive integer, then
Γ = (n − 1)!
showing the connection to the factorial function.

Inverse

In mathematics, the inverse of a function y = f(x) is a function that, in some fashion, "undoes" the effect of f. The inverse of f is denoted f^{-1}. The statements y=f(x) and x=f^{-1}(y) are equivalent.

Inverse Laplace transform

In mathematics, the Inverse Laplace transform of F(s) is the function f(t) which has the property

Chapter 7. THE LAPLACE TRANSFORM

Chapter 7. THE LAPLACE TRANSFORM

$$\mathcal{L}\{f(t)\} = F(s),$$

where $\mathcal{L}$ is the Laplace transform.

It can be proven, that if a function F(s) has the Inverse Laplace transform f(t), i.e. f is a piecewise continuous and exponentially restricted real function f satisfying the condition

$$\mathcal{L}\{f(t)\} = F(s)$$

then f(t) is uniquely determined.

The Laplace transform and the Inverse Laplace transform together have a number of properties that make them useful for analysing linear dynamic systems.

Partial fraction	In algebra, the Partial fraction decomposition or Partial fraction expansion is used to reduce the degree of either the numerator or the denominator of a rational function. The outcome of a full Partial fraction expansion expresses that function as a sum of fractions, where: · the denominator of each term is a power of an irreducible (not factorable) polynomial and · the numerator is a polynomial of smaller degree than that irreducible polynomial. In symbols, one can use Partial fraction expansion to change a rational function in the form $$\frac{f(x)}{g(x)}$$ where f and g are polynomials, into a function of the form $$\sum \frac{a_n}{h_n(x)}$$ where h_n are polynomials that are factors of g(x), and are in general of lower degree.
Partial fractions	In integral calculus we would want to write a fractional algebraic expression as the sum of its partial fractions in order to take the integral of each simple fraction separately. Once the original denominator, D_0, has been factored we set up a fraction for each factor in the denominator. We may use a subscripted D to represent the denominator of the respective partial fractions which are the factors in D_0.
Derivative	In calculus (a branch of mathematics) the Derivative is a measure of how a function changes as its input changes. Loosely speaking, a Derivative can be thought of as how much a quantity is changing at a given point; for example, the Derivative of the position of a vehicle with respect to time is the instantaneous velocity at which the vehicle is traveling. Conversely, the integral of the velocity over time is the change in the vehicle"s position.
Initial value problem	In mathematics, in the field of differential equations, an initial value problem is an ordinary differential equation together with specified value, called the initial condition, of the unknown function at a given point in the domain of the solution. In physics or other sciences, modeling a system frequently amounts to solving an initial value problem; in this context, the differential equation is an evolution equation specifying how, given initial conditions, the system will evolve with time. An initial value problem is a differential equation

Chapter 7. THE LAPLACE TRANSFORM

Chapter 7. THE LAPLACE TRANSFORM

$$y'(t) = f(t, y(t)) \quad \text{with} \quad f : \mathbb{R} \times \mathbb{R} \to \mathbb{R}$$

together with a point in the domain of f

$$(t_0, y_0) \in \mathbb{R} \times \mathbb{R},$$

called the initial condition.

Translation	Translation is the interpreting of the meaning of a text and the subsequent production of an equivalent text, likewise called a Translation " that communicates the same message in another language. The text to be translated is called the source text, and the language that it is to be translated into is called the target language; the final product is sometimes called the target text.
Theorem	In mathematics, a Theorem is a statement proved on the basis of previously accepted or established statements such as axioms. In formal mathematical logic, the concept of a Theorem may be taken to mean a formula that can be derived according to the derivation rules of a fixed formal system. The statements of a theory as expressed in a formal language are called its elementary Theorem s and are said to be true.
Unit	In mathematics, a Unit in a (unital) ring R is an invertible element of R, i.e. an element u such that there is a v in R with uv = vu = 1_R, where 1_R is the multiplicative identity element. That is, u is an invertible element of the multiplicative monoid of R. If $0 \neq 1$ in the ring, then 0 is not a Unit. Unfortunately, the term Unit is also used to refer to the identity element 1_R of the ring, in expressions like ring with a Unit or Unit ring, and also e.g. "Unit" matrix.
Heaviside step function	The Heaviside step function, H is a discontinuous function whose value is zero for negative argument and one for positive argument. It seldom matters what value is used for H(0), since H is mostly used as a distribution. Some common choices can be seen below.
Second	The second (SI symbol: s), sometimes abbreviated sec., is the name of a unit of time, and is the International System of Units (SI) base unit of time. It may be measured using a clock. SI prefixes are frequently combined with the word second to denote subdivisions of the second e.g., the milli second (one thousandth of a second , the micro second (one millionth of a second , and the nano second (one billionth of a second)
Convolution	In mathematics and, in particular, functional analysis, Convolution is a mathematical operation on two functions f and g, producing a third function that is typically viewed as a modified version of one of the original functions. Convolution is similar to cross-correlation. It has applications that include statistics, computer vision, image and signal processing, electrical engineering, and differential equations.
Convolution theorem	In mathematics, the Convolution theorem states that under suitable conditions the Fourier transform of a convolution is the pointwise product of Fourier transforms. In other words, convolution in one domain (e.g., time domain) equals point-wise multiplication in the other domain (e.g., frequency domain). Versions of the Convolution theorem are true for various Fourier-related transforms.

Chapter 7. THE LAPLACE TRANSFORM

Chapter 7. THE LAPLACE TRANSFORM

Commutative property	Records of the implicit use of the commutative property go back to ancient times. The Egyptians used the commutative property of multiplication to simplify computing products. Euclid is known to have assumed the commutative property of multiplication in his book Elements.
Integral equation	In mathematics, an Integral equation is an equation in which an unknown function appears under an integral sign. There is a close connection between differential and Integral equations, and some problems may be formulated either way. See, for example, Maxwell"s equations.
Voltage	voltage is commonly used as a short name for electrical potential difference. Its corresponding SI unit is the volt (symbol: V, not italicized). Electric potential is a hypothetically measurable physical dimension, and is denoted by the algebraic variable V (italicized).
Voltage drop	Voltage drop is the reduction in voltage in an electrical circuit between the source and load. In electrical wiring national and local electrical codes may set guidelines for maximum Voltage drop allowed in a circuit, to ensure reasonable efficiency of distribution and proper operation of electrical equipment . Voltage drop may be neglected when the impedance of the interconnecting conductors is small relative to the other components of the circuit.
Volterra integral equation	In mathematics, the Volterra integral equations are a special type of integral equations. They are divided into two groups referred to as the first and the second kind. A linear Volterra equation of the first kind is $$f(t) = \int_a^t K(t,s)\, x(s)\, ds.$$ A linear Volterra equation of the second kind is $$x(t) = f(t) + \int_a^t K(t,s) x(s)\, ds.$$ In operator theory, and in Fredholm theory, the corresponding equations are called the Volterra operator.
Periodic function	In mathematics, a Periodic function is a function that repeats its values in regular intervals or periods. The most important examples are the trigonometric functions, which repeat over intervals of length 2π. Periodic function s are used throughout science to describe oscillations, waves, and other phenomena that exhibit periodicity.
Square wave	A Square wave is a kind of non-sinusoidal waveform, most typically encountered in electronics and signal processing. An ideal Square wave alternates regularly and instantaneously between two levels. Square waves are universally encountered in digital switching circuits and are naturally generated by binary (two-level) logic devices.
Meander	A Meander in general is a bend in a sinuous watercourse. A Meander is formed when the moving water in a river erodes the outer banks and widens its valley. A stream of any volume may assume a Meandering course, alternatively eroding sediments from the outside of a bend and depositing them on the inside.
Laguerre polynomials	In mathematics, the Laguerre polynomials are the canonical solutions of Laguerre"s equation:

Chapter 7. THE LAPLACE TRANSFORM

Chapter 7. THE LAPLACE TRANSFORM

$$x y'' + (1 - x) y' + n y = 0$$

which is a second-order linear differential equation. This equation has nonsingular solutions only if n is a non-negative integer. The Laguerre polynomials are also used for Gaussian quadrature to numerically compute integrals of the form

$$\int_0^\infty f(x)dx$$

Differential equation

A Differential equation is a mathematical equation for an unknown function of one or several variables that relates the values of the function itself and its derivatives of various orders. Differential equations play a prominent role in engineering, physics, economics and other disciplines. Visualization of airflow into a duct modelled using the Navier-Stokes equations, a set of partial Differential equations.

Differential equations arise in many areas of science and technology: whenever a deterministic relationship involving some continuously changing quantities (modelled by functions) and their rates of change (expressed as derivatives) is known or postulated.

Rectification

Rectification has the following technical meanings:

· Rectification, in astrology
· Rectification (chemistry), a concept found in biology and industrial chemistry
· Chinese history: see Cheng Feng
· Chinese philosophy: see Confucianism f of names
· Rectification (geometry) - truncating a polytope by marking the midpoints of all its edges, and cutting off its vertices at those points
· Rectifier (electricity) - converter of AC to DC
· Image Rectification - adjustment of images to simplify stereo vision or to map images to a map coordinate system (GIS)
· Rectification (law)
· Rectifiable curve, in mathematics
· Rectifiable set, in mathematics ".

Sine

The sine of an angle is the ratio of the length of the opposite side to the length of the hypotenuse. In our case

$$\sin A = \frac{\text{opposite}}{\text{hypotenuse}} = \frac{a}{h}.$$

Note that this ratio does not depend on size of the particular right triangle chosen, as long as it contains the angle A, since all such triangles are similar.

The co sine of an angle is the ratio of the length of the adjacent side to the length of the hypotenuse.

Sine wave

The Sine wave or sinusoid is a function that occurs often in mathematics, music, physics, signal processing, audition, electrical engineering, and many other fields. Its most basic form is:

$$y(t) = A \cdot \sin(\omega t + \theta)$$

which describes a wavelike function of time (t) with:

Chapter 7. THE LAPLACE TRANSFORM

Chapter 7. THE LAPLACE TRANSFORM

- peak deviation from center = A (aka amplitude)
- angular frequency ω, (radians per second)
- phase = θ

- When the phase is non-zero, the entire waveform appears to be shifted in time by the amount θ/ω seconds. A negative value represents a delay, and a positive value represents a "head-start".

Problems listening to this file? See media help.

The Sine wave is important in physics because it retains its waveshape when added to another Sine wave of the same frequency and arbitrary phase. It is the only periodic waveform that has this property. This property leads to its importance in Fourier analysis and makes it acoustically unique.

Dirac delta	The Dirac delta or Dirac"s delta is a mathematical construct introduced by theoretical physicist Paul Dirac. Informally, it is a generalized function representing an infinitely sharp peak bounding unit area: a "function" δ(x) that has the value zero everywhere except at x = 0 where its value is infinitely large in such a way that its total integral is 1. In the context of signal processing it is often referred to as the unit impulse function.
	The Dirac delta is not strictly a function, because any function that is equal to zero everywhere but a single point must have total integral zero.
Unit impulse	The Dirac delta or Dirac"s delta is a mathematical construct introduced by theoretical physicist Paul Dirac. Informally, it is a generalized function representing an infinitely sharp peak bounding unit area: a "function" δ(x) that has the value zero everywhere except at x = 0 where its value is infinitely large in such a way that its total integral is 1. In the context of signal processing it is often referred to as the unit impulse function.
	The Dirac delta is not strictly a function, because any function that is equal to zero everywhere but a single point must have total integral zero.
Distributions	In mathematical analysis, distributions (or generalized functions) are objects that generalize functions. distributions make it possible to differentiate functions whose derivative does not exist in the classical sense. In particular, any locally integrable function has a distributional derivative.
Generalized functions	In mathematics, Generalized functions are objects generalizing the notion of functions. There is more than one recognised theory. Generalized functions are especially useful in making discontinuous functions more like smooth functions, and (going to extremes) describing physical phenomena such as point charges.
Theory	The term Theory has two broad sets of meanings, one used in the empirical sciences (both natural and social) and the other used in philosophy, mathematics, logic, and across other fields in the humanities. There is considerable difference and even dispute across academic disciplines as to the proper usages of the term. What follows is an attempt to describe how the term is used, not to try to say how it ought to be used.

Chapter 7. THE LAPLACE TRANSFORM

Chapter 7. THE LAPLACE TRANSFORM

Weight function

A Weight function is a mathematical device used when performing a sum, integral, or average in order to give some elements more "weight" or influence on the result than other elements in the same set. They occur frequently in statistics and analysis, and are closely related to the concept of a measure. Weight function s can be employed in both discrete and continuous settings.

Linear system

A linear system is a mathematical model of a system based on the use of a linear operator. linear systems typically exhibit features and properties that are much simpler than the general, nonlinear case. As a mathematical abstraction or idealization, linear systems find important applications in automatic control theory, signal processing, and telecommunications.

Mathematical model

Note: The term model has a different meaning in model theory, a branch of mathematical logic. An artifact which is used to illustrate a mathematical idea is also called a Mathematical model and this usage is the reverse of the sense explained below.

A Mathematical model uses mathematical language to describe a system.

Linear differential equation

In mathematics, a Linear differential equation is of the form

$$Ly = f$$

where the differential operator L is a linear operator, y is the unknown function (such as a function of time y(t)), and the right hand side f is a given function of the same nature as y (called the source term). For a function dependent on time we may write the equation more expressively as

$$Ly(t) = f(t)$$

and, even more precisely by bracketing

$$L[y(t)] = f(t)$$

The linear operator L may be considered to be of the form

$$L_n(y) \equiv \frac{d^n y}{dt^n} + A_1(t)\frac{d^{n-1}y}{dt^{n-1}} + \cdots + A_{n-1}(t)\frac{dy}{dt} + A_n(t)y$$

The linearity condition on L rules out operations such as taking the square of the derivative of y; but permits, for example, taking the second derivative of y. It is convenient to rewrite this equation in an operator form

$$L_n(y) \equiv \left[D^n + A_1(t)D^{n-1} + \cdots + A_{n-1}(t)D + A_n(t) \right] y$$

where D is the differential operator d/dt (i.e. Dy = y", D^2y = y",...

Linear model

In statistics, the term linear model is used in different ways according to the context. The most common occurrence is in connection with regression models and the term is often taken as synonymous with linear regression model. However the term is also used in time series analysis with a different meaning.

Ordinary differential equation

In mathematics, an ordinary differential equation (or ordinary differential equation) is a relation that contains functions of only one independent variable, and one or more of its derivatives with respect to that variable.

Chapter 7. THE LAPLACE TRANSFORM

Chapter 7. THE LAPLACE TRANSFORM

A simple example is Newton's second law of motion, which leads to the differential equation

$$m\frac{d^2x(t)}{dt^2} = F(x(t)),$$

for the motion of a particle of constant mass m. In general, the force F depends upon the position of the particle x(t) at time t, and thus the unknown function x(t) appears on both sides of the differential equation, as is indicated in the notation F(x(t).)

Spring

In geometry, a Spring is a surface of revolution in the shape of a helix with thickness, generated by revolving a circle about the path of a helix. The torus is a special case of the Spring obtained when the helix is crushed to a circle.

A Spring wrapped around the z-axis can be defined parametrically by:

$$x(u,v) = (R + r\cos v)\cos u,$$
$$y(u,v) = (R + r\cos v)\sin u,$$
$$z(u,v) = r\sin v + \frac{P\cdot u}{\pi},$$

where

$$u \in [0,\ 2n\pi]\ (n \in \mathbb{R}),$$
$$v \in [0,\ 2\pi],$$

R is the distance from the center of the tube to the center of the helix,
r is the radius of the tube,
P is the speed of the movement along the z axis (in a right-handed Cartesian coordinate system, positive values create right-handed springs, whereas negative values create left-handed springs),
n is the number of rounds in circle.

Lissajous curve

In mathematics, a Lissajous curve is the graph of the system of parametric equations

$$x = A\sin(at + \delta),\quad y = B\sin(bt),$$

which describes complex harmonic motion. This family of curves was investigated by Nathaniel Bowditch in 1815, and later in more detail by Jules Antoine Lissajous in 1857.

The appearance of the figure is highly sensitive to the ratio a/b.

Chapter 7. THE LAPLACE TRANSFORM

Chapter 8. SYSTEMS OF LINEAR FIRST-ORDER DIFFERENTIAL EQUATIONS

Differential equation

A Differential equation is a mathematical equation for an unknown function of one or several variables that relates the values of the function itself and its derivatives of various orders. Differential equations play a prominent role in engineering, physics, economics and other disciplines. Visualization of airflow into a duct modelled using the Navier-Stokes equations, a set of partial Differential equations.

Differential equations arise in many areas of science and technology: whenever a deterministic relationship involving some continuously changing quantities (modelled by functions) and their rates of change (expressed as derivatives) is known or postulated.

Ordinary differential equation

In mathematics, an ordinary differential equation (or ordinary differential equation) is a relation that contains functions of only one independent variable, and one or more of its derivatives with respect to that variable.

A simple example is Newton"s second law of motion, which leads to the differential equation

$$m\frac{d^2x(t)}{dt^2} = F(x(t)),$$

for the motion of a particle of constant mass m. In general, the force F depends upon the position of the particle x(t) at time t, and thus the unknown function x(t) appears on both sides of the differential equation, as is indicated in the notation F(x(t).)

Coefficient matrix

In linear algebra, the Coefficient matrix refers to a matrix consisting of the coefficients of the variables in a set of linear equations.

In general, a system with m linear equations and n unknowns can be written as

$$a_{11}x_1 + a_{12}x_2 + ... + a_{1n}x_n = b_1$$
$$a_{21}x_1 + a_{22}x_2 + ... + a_{2n}x_n = b_2$$
$$\vdots$$
$$a_{m1}x_1 + a_{m2}x_2 + ... + a_{mn}x_n = b_m$$

where $x_1, x_2, ..., x_n$ are the unknowns and the numbers $a_{11}, a_{12}, ..., a_{mn}$ are the coefficients of the system. The Coefficient matrix is the mxn matrix with the coefficient a_{ij} as the (i,j)-th entry:

$$\begin{bmatrix} a_{11} & a_{12} & \cdots & a_{1n} \\ a_{21} & a_{22} & \cdots & a_{2n} \\ \vdots & \vdots & \ddots & \vdots \\ a_{m1} & a_{m2} & \cdots & a_{mn} \end{bmatrix}$$

Homogeneous function

In mathematics, a Homogeneous function is a function with multiplicative scaling behaviour: if the argument is multiplied by a factor, then the result is multiplied by some power of this factor.

Suppose that $f : V \to W$ is a function between two vector spaces over a field F.

We say that f is homogeneous of degree k if

$$f(\alpha \mathbf{v}) = \alpha^k f(\mathbf{v})$$

Chapter 8. SYSTEMS OF LINEAR FIRST-ORDER DIFFERENTIAL EQUATIONS

Chapter 8. SYSTEMS OF LINEAR FIRST-ORDER DIFFERENTIAL EQUATIONS

for all nonzero $\alpha \in F$ and $\mathbf{v} \in V$.

Linear	In a different usage to the above, a polynomial of degree 1 is said to be Linear, because the graph of a function of that form is a line.
	Over the reals, a Linear equation is one of the form:
	f(x) = m x + b
	m is often called the slope or gradient; b the y-intercept, which gives the point of intersection between the graph of the function and the y-axis.
	Note that this usage of the term Linear is not the same as the above, because Linear polynomials over the real numbers do not in general satisfy either additivity or homogeneity.
Linear system	A linear system is a mathematical model of a system based on the use of a linear operator. linear systems typically exhibit features and properties that are much simpler than the general, nonlinear case. As a mathematical abstraction or idealization, linear systems find important applications in automatic control theory, signal processing, and telecommunications.
Matrix	In mathematics, a matrix is a rectangular array of numbers, such as $$\begin{bmatrix} 1 & 2 & 3 \\ 6 & 5 & 4 \end{bmatrix}.$$ Entries of a matrix are often denoted by a variable with two subscripts, as shown on the right. Matrices of the same size can be added and subtracted entrywise and matrices of compatible size can be multiplied. These operations have many of the properties of ordinary arithmetic, except that matrix multiplication is not commutative, that is, AB and BA are not equal in general.
Normal	In abstract algebra, an algebraic field extension L/K is said to be normal if L is the splitting field of a family of polynomials in K[X]. Bourbaki calls such an extension a quasi-Galois extension.
	The normality of L/K is equivalent to each of the following properties:
	· Let K^a be an algebraic closure of K containing L. Every embedding σ of L in K^a which restricts to the identity on K, satisfies σ(L) = L. In other words, σ is an automorphism of L over K.
	· Every irreducible polynomial in K[X] which has a root in L factors into linear factors in L[X].
	For example, $\mathbb{Q}(\sqrt{2})$ is a normal extension of $\mathbb{Q}$, since it is the splitting field of $x^2 - 2$. On the other hand, $\mathbb{Q}(\sqrt[3]{2})$ is not a normal extension of $\mathbb{Q}$ since the polynomial $x^3 - 2$ has one root in it (namely, $\sqrt[3]{2}$), but not all of them (it does not have the non-real cubic roots of 2).
Normal form	In game theory, normal form is a way of describing a game. Unlike extensive form, normal-form representations are not graphical per se, but rather represent the game by way of a matrix. While this approach can be of greater use in identifying strictly dominated strategies and Nash equilibria, some information is lost as compared to extensive-form representations.

Chapter 8. SYSTEMS OF LINEAR FIRST-ORDER DIFFERENTIAL EQUATIONS

Chapter 8. SYSTEMS OF LINEAR FIRST-ORDER DIFFERENTIAL EQUATIONS

Degree

In topology, the term degree is applied to continuous maps between manifolds of the same dimension. The degree of a map can be defined in terms of homology groups or, for smooth maps, in terms of preimages of regular values. It is a generalization of winding number.

Phase plane

A Phase plane is a visual display of certain characteristics of certain kinds of differential equations.

Phase plane s are useful in visualizing the behavior of physical systems; in particular, of oscillatory systems such as predator-prey models These models can "spiral in" towards zero, "spiral out" towards infinity, or reach neutrally stable situations called centres where the path traced out can be either circular, elliptical, or ovoid, or some variant thereof.

Variation of parameters

In mathematics, variation of parameters also known as variation of constants, is a general method to solve inhomogeneous linear ordinary differential equations. It was developed by the Italian-French mathematician Joseph Louis Lagrange with noteworthy help from the American mathematician and physicist Noah LaMoyne.

For first-order inhomogeneous linear differential equations it"s usually possible to find solutions via integrating factors or undetermined coefficients with considerably less effort, although those methods are rather heuristics that involve guessing and don"t work for all inhomogenous linear differential equations.

Linear differential equation

In mathematics, a Linear differential equation is of the form

$$Ly = f$$

where the differential operator L is a linear operator, y is the unknown function (such as a function of time y(t)), and the right hand side f is a given function of the same nature as y (called the source term). For a function dependent on time we may write the equation more expressively as

$$Ly(t) = f(t)$$

and, even more precisely by bracketing

$$L[y(t)] = f(t)$$

The linear operator L may be considered to be of the form

$$L_n(y) \equiv \frac{d^n y}{dt^n} + A_1(t)\frac{d^{n-1} y}{dt^{n-1}} + \cdots + A_{n-1}(t)\frac{dy}{dt} + A_n(t)y$$

The linearity condition on L rules out operations such as taking the square of the derivative of y; but permits, for example, taking the second derivative of y. It is convenient to rewrite this equation in an operator form

$$L_n(y) \equiv \left[D^n + A_1(t)D^{n-1} + \cdots + A_{n-1}(t)D + A_n(t) \right] y$$

where D is the differential operator d/dt (i.e. Dy = y" , D^2y = y",...

Orthogonal

In mathematics, two vectors are orthogonal if they are perpendicular, i.e., they form a right angle. The word comes from the Greek á½€ρθÏŒς , meaning "straight", and γωνῖα (gonia), meaning "angle". For example, a subway and the street above, although they do not physically intersect, are orthogonal if they cross at a right angle.

Parameter

In mathematics, statistics, and the mathematical sciences, a Parameter is a quantity that defines certain characteristics of systems or functions that serves to relate functions and variables using a common variable (often t) when such a relationship would be difficult to explicate with an equation. In different contexts the term may have special uses.

Chapter 8. SYSTEMS OF LINEAR FIRST-ORDER DIFFERENTIAL EQUATIONS

· In a section on frequently misused words in his book The Writer"s Art, James J. Kilpatrick quoted a letter from a correspondent, giving examples to illustrate the correct use of the word Parameter

· A parametric equaliser is an audio filter that allows the frequency of maximum cut or boost to be set by one control, and the size of the cut or boost by another.

Parametric equations	In mathematics, Parametric equations are a method of defining a function using parameters. A simple kinematical example is when one uses a time parameter to determine the position, velocity, and other information about a body in motion.
	Abstractly, a relation is given in the form of an equation, and it is shown also to be the image of functions from items such as R^n.
Vector	In elementary mathematics, physics, and engineering, a vector is a geometric object that has both a magnitude (or length), direction and sense, (i.e., orientation along the given direction.) A vector is frequently represented by a line segment with a definite direction, or graphically as an arrow, connecting an initial point A with a terminal point B, and denoted by $\overrightarrow{AB}$.
	The magnitude of the vector is the length of the segment and the direction characterizes the displacement of B relative to A: how much one should move the point A to "carry" it to the point B.
	Many algebraic operations on real numbers have close analogues for vectors.
Initial value problem	In mathematics, in the field of differential equations, an initial value problem is an ordinary differential equation together with specified value, called the initial condition, of the unknown function at a given point in the domain of the solution. In physics or other sciences, modeling a system frequently amounts to solving an initial value problem; in this context, the differential equation is an evolution equation specifying how, given initial conditions, the system will evolve with time.
	An initial value problem is a differential equation $$y'(t) = f(t, y(t)) \quad \text{with} \quad f : \mathbb{R} \times \mathbb{R} \to \mathbb{R}$$ together with a point in the domain of f $$(t_0, y_0) \in \mathbb{R} \times \mathbb{R},$$ called the initial condition.
Interval	In mathematics, a (real) interval is a set of real numbers with the property that any number that lies between two numbers in the set is also included in the set. For example, the set of all numbers x satisfying $0 \leq x \leq 1$ is an interval which contains 0 and 1, as well as all numbers between them. Other examples of intervals are the set of all real numbers $\mathbb{R}$, the set of all negative real numbers, and the empty set.
Superposition principle	In physics and systems theory, the Superposition principle also known as superposition property, states that, for all linear systems,
	> The net response at a given place and time caused by two or more stimuli is the sum of the responses which would have been caused by each stimulus individually.

Chapter 8. SYSTEMS OF LINEAR FIRST-ORDER DIFFERENTIAL EQUATIONS

Chapter 8. SYSTEMS OF LINEAR FIRST-ORDER DIFFERENTIAL EQUATIONS

So that if input A produces response X and input B produces response Y then input (A + B) produces response (X + Y.) Mathematically, for all linear systems F(x) = y, where x is some sort of stimulus (input) and y is some sort of response (output), the superposition (i.e., sum) of stimuli yields a superposition of the respective responses:

$$F(x_1 + x_2 + \cdots) = F(x_1) + F(x_2) + \cdots$$

Uniqueness

In mathematics and logic, the phrase "there is one and only one" is used to indicate that exactly one object with a certain property exists. In mathematical logic, this sort of quantification is known as Uniqueness quantification or unique existential quantification.

Uniqueness quantification is often denoted with the symbols "∃!" or $\exists_{=1}$".

Theorem

In mathematics, a Theorem is a statement proved on the basis of previously accepted or established statements such as axioms. In formal mathematical logic, the concept of a Theorem may be taken to mean a formula that can be derived according to the derivation rules of a fixed formal system. The statements of a theory as expressed in a formal language are called its elementary Theorem s and are said to be true.

Wronskian

In mathematics, the Wronskian is a function especially important in the study of differential equations, where it can be used to determine whether a set of solutions is linearly independent.

For n real- or complex-valued functions f_1, ..., f_n, which are n − 1 times differentiable on an interval I, the Wronskian $W(f_1, ..., f_n)$ as a function on I is defined by

$$W(f_1, \ldots, f_n)(x) = \begin{vmatrix} f_1(x) & f_2(x) & \cdots & f_n(x) \\ f_1'(x) & f_2'(x) & \cdots & f_n'(x) \\ \vdots & \vdots & \ddots & \vdots \\ f_1^{(n-1)}(x) & f_2^{(n-1)}(x) & \cdots & f_n^{(n-1)}(x) \end{vmatrix}, \quad x \in I.$$

That is, it is the determinant of the matrix constructed by placing the functions in the first row, the first derivative of each function in the second row, and so on through the (n - 1)st derivative, thus forming a square matrix sometimes called a fundamental matrix.

Differential operator

In mathematics, a Differential operator is an operator defined as a function of the differentiation operator. It is helpful, as a matter of notation first, to consider differentiation as an abstract operation, accepting a function and returning another (in the style of a higher-order function in computer science).

There are certainly reasons not to restrict to linear operators; for instance the Schwarzian derivative is a well-known non-linear operator.

Nominative determinism

Nominative determinism refers to the theory that a person"s name is given an influential role in reflecting key attributes of his job, profession, but real examples are more highly prized, the more obscure the better.

Chapter 8. SYSTEMS OF LINEAR FIRST-ORDER DIFFERENTIAL EQUATIONS

Chapter 8. SYSTEMS OF LINEAR FIRST-ORDER DIFFERENTIAL EQUATIONS

Weak solution	In mathematics, a weak solution to an ordinary or partial differential equation is a function for which the derivatives appearing in the equation may not all exist but which is nonetheless deemed to satisfy the equation in some precisely defined sense. There are many different definitions of weak solution, appropriate for different classes of equations. One of the most important is based on the notion of distributions.
Defined	In mathematics, defined and undefined are used to explain whether or not expressions have meaningful, sensible, and unambiguous values. Whether an expression has a meaningful value depends on the context of the expression. For example the value of 4 − 5 is undefined if an positive integer result is required.
Laplace transform	In mathematics, the Laplace transform is a widely used integral transform. It has many important applications in mathematics, physics, optics, electrical engineering, control engineering, signal processing, and probability theory. The Laplace transform is related to the Fourier transform, but whereas the Fourier transform resolves a function or signal into its modes of vibration, the Laplace transform resolves a function into its moments.
Eigenvalue	In mathematics, eigenvalue, eigenvector, and eigenspace are related concepts in the field of linear algebra. Linear algebra studies linear transformations, which are represented by matrices acting on vectors. eigenvalues, eigenvectors and eigenspaces are properties of a matrix.
Mathieu functions	In mathematics, the Mathieu functions are certain special functions useful for treating a variety of problems in applied mathematics, including · vibrating elliptical drumheads, · quadrupoles mass filters and quadrupole ion traps for mass spectrometry · the phenomenon of parametric resonance in forced oscillators, · exact plane wave solutions in general relativity. · the Stark effect for a rotating electric dipole. They were introduced by Émile Léonard Mathieu in 1868 in the context of the first problem. The canonical form for Mathieu"s differential equation is $$\frac{d^2u}{dx^2} + [a_u - 2q_u \cos(2x)]u = 0.$$ Closely related is Mathieu"s modified differential equation $$\frac{d^2y}{du^2} - [a - 2q\cosh(2u)]y = 0$$ which follows on substitution u = ix.
Distinct	Two or more things are Distinct if no two of them are the same thing. In mathematics, two things are called Distinct if they are not equal. A quadratic equation over the complex numbers sometimes has two roots.
Eigenvector	In mathematics, eigenvalue, eigenvector, and eigenspace are related concepts in the field of linear algebra. Linear algebra studies linear transformations, which are represented by matrices acting on vectors. Eigenvalues, eigenvectors and eigenspaces are properties of a matrix.

Chapter 8. SYSTEMS OF LINEAR FIRST-ORDER DIFFERENTIAL EQUATIONS

Chapter 8. SYSTEMS OF LINEAR FIRST-ORDER DIFFERENTIAL EQUATIONS

Real	A point (a,b,c) in the complex projective plane is called real if there exists a complex number z such that za, zb and zc are all real numbers.
	This definition can be widened to complex projective space and complex projective hyperspaces as follows:
	$(a_1, a_2, ..., a_n)$
	is real if there exists a complex number z such that
	$(za_1, za_2, ..., za_n)$
	is real.
	(Note (0,0,...,0) is not a point)
Phase portrait	A Phase portrait is a geometric representation of the trajectories of a dynamical system in the phase plane. Each set of initial conditions is representated by a different curve, or point.
	Phase portrait s are an invaluable tool in studying dynamical systems.
Attractor	An attractor is a set to which a dynamical system evolves after a long enough time. That is, points that get close enough to the attractor remain close even if slightly disturbed. Geometrically, an attractor can be a point, a curve, a manifold, or even a complicated set with a fractal structure known as a strange attractor.
Multiplicity	In mathematics, the multiplicity of a member of a multiset is how many memberships in the multiset it has. For example, the term is used to refer to the number of times a given polynomial equation has a root at a given point.
	The common reason to consider notions of multiplicity is to count correctly, without specifying exceptions (for example, double roots counted twice).
Symmetry group	The Symmetry group is sometimes also called full Symmetry group in order to emphasize that it includes the orientation-reversing isometries under which the figure is invariant. The subgroup of orientation-preserving isometries which leave the figure invariant is called its proper Symmetry group The proper Symmetry group of an object is equal to its full Symmetry group if and only if the object is chiral
Symmetric matrix	In linear algebra, a Symmetric matrix is a square matrix, A, that is equal to its transpose
	$A = A^T.$
	The entries of a Symmetric matrix are symmetric with respect to the main diagonal (top left to bottom right.) So if the entries are written as A = (a_{ij}), then
	$a_{ij} = a_{ji}$

Chapter 8. SYSTEMS OF LINEAR FIRST-ORDER DIFFERENTIAL EQUATIONS

Chapter 8. SYSTEMS OF LINEAR FIRST-ORDER DIFFERENTIAL EQUATIONS

for all indices i and j. The following 3×3 matrix is symmetric:

$$\begin{bmatrix} 1 & 2 & 3 \\ 2 & 4 & -5 \\ 3 & -5 & 6 \end{bmatrix}.$$

A matrix is called skew-symmetric or antisymmetric if its transpose is the same as its negative.

Mass matrix

In computational mechanics, a Mass matrix is a generalization of the concept of mass to generalized coordinates. For example, consider a two-body particle system in one dimension. The position of such a system has two degrees of freedom, the position of each particle, which can be described by the generalized position vector

$$\mathbf{x} = \begin{bmatrix} x_1 & x_2 \end{bmatrix}^\top.$$

Hyperbolic cosine

In mathematics, the hyperbolic functions are analogs of the ordinary trigonometric, functions. The basic hyperbolic functions are the hyperbolic sine "sinh", and the hyperbolic cosine "cosh", from which are derived the hyperbolic tangent "tanh", etc., in analogy to the derived trigonometric functions. The inverse hyperbolic functions are the area hyperbolic sine "arsinh" (also called "asinh", or sometimes by the misnomer of "arcsinh") and so on.

Frequency

Frequency is the number of occurrences of a repeating event per unit time. It is also referred to as temporal Frequency. The period is the duration of one cycle in a repeating event, so the period is the reciprocal of the Frequency.

Method of undetermined coefficients

In mathematics, the method of undetermined coefficients is an approach to finding a particular solution to certain inhomogeneous ordinary differential equations and recurrence relations. It is closely related to the annihilator method, but instead of using a particular kind of differential operator (the annihilator) in order to find the best possible form of the particular solution, a "guess" is made as to the appropriate form, which is then tested by differentiating the resulting equation. In this sense, the method of undetermined coefficients is less formal but more intuitive than the annihilator method.

Matrix derivative

In mathematics, matrix calculus is a specialized notation for doing multivariable calculus, especially over spaces of matrices, where it defines the matrix derivative. This notation is well-suited to describing systems of differential equations, and taking derivatives of matrix-valued functions with respect to matrix variables. This notation is commonly used in statistics and engineering, while the tensor index notation is preferred in physics.

Definition

A Definition is a formal passage describing the meaning of a term (a word or phrase). The term to be defined is the definiendum . A term may have many subtly different senses or meanings.

Nilpotent

In mathematics, an element x of a ring R is called Nilpotent if there exists some positive integer n such that $x^n = 0$.

The term was introduced by Benjamin Peirce in the context of elements of an algebra that vanish when raised to a power.

· This definition can be applied in particular to square matrices. The matrix

Chapter 8. SYSTEMS OF LINEAR FIRST-ORDER DIFFERENTIAL EQUATIONS

$$A = \begin{pmatrix} 0 & 1 & 0 \\ 0 & 0 & 1 \\ 0 & 0 & 0 \end{pmatrix}$$

is Nilpotent because $A^3 = 0$.

Nilpotent matrix

In linear algebra, a Nilpotent matrix is a square matrix N such that

$$N^k = 0$$

for some positive integer k. The smallest such k is sometimes called the degree of N.

More generally, a nilpotent transformation is a linear transformation L of a vector space such that $L^k = 0$ for some positive integer k. Both of these concepts are special cases of a more general concept of nilpotence that applies to elements of rings.

Chapter 8. SYSTEMS OF LINEAR FIRST-ORDER DIFFERENTIAL EQUATIONS

Chapter 9. NUMERICAL SOLUTIONS OF ORDINARY DIFFERENTIAL EQUATIONS

Round-off error

A Round-off error is the difference between the calculated approximation of a number and its exact mathematical value. Numerical analysis specifically tries to estimate this error when using approximation equations and/or algorithms, especially when using finite digits to represent real numbers This is a form of quantization error.

Discretization error

In numerical analysis, computational physics, and simulation, Discretization error is error resulting from the fact that a function of a continuous variable is represented in the computer by a finite number of evaluations, for example, on a lattice. Discretization error can usually be reduced by using a more finely spaced lattice, with an increased computational cost.

In signal processing, the analog of discretization is sampling, and results in no loss if the conditions of the sampling theorem are satisfied, otherwise the resulting error is called aliasing.

Formula

In mathematics, a Formula is a key to solve an equation with variables. For example, the problem of determining the volume of a sphere is one that requires a significant amount of integral calculus to solve. However, having done this once, mathematicians can produce a Formula to describe the volume in terms of some other parameter (the radius for example).

Truncation

In mathematics, Truncation is the term for limiting the number of digits right of the decimal point, by discarding the least significant ones.

For example, consider the real numbers

5.6341432543653654
32.438191288
-6.3444444444444

To truncate these numbers to 4 decimal digits, we only consider the 4 digits to the right of the decimal point.
The result would be:

5.6341
32.4381
-6.3444

Note that in some cases, truncating would yield the same result as rounding, but Truncation does not round up or round down the digits; it merely cuts off at the specified digit.

Truncation error

Truncation error or local Truncation error is error made by numerical algorithms that arises from taking finite number of steps in computation. It is present even with infinite-precision arithmetic, because it is caused by truncation of the infinite Taylor series to form the algorithm.

Use of arbitrarily small steps in numerical computation is prevented by round-off error, which are the consequence of using finite precision floating point numbers on computers.

Order

Order refers to a large number of concepts in mathematics, especially in algebra, arithmetic, analysis, combinatorics, fractals, graphs, and mathematical theories.

Chapter 9. NUMERICAL SOLUTIONS OF ORDINARY DIFFERENTIAL EQUATIONS

Chapter 9. NUMERICAL SOLUTIONS OF ORDINARY DIFFERENTIAL EQUATIONS

- Order of computation, the computational complexity of an algorithm
- Canonical Order, the Order of elements that obeys a certain set of rules or specifications
- Z-Order, the Order of windows on computer screens

- First-Order hold in signal processing
- Modulation Order, the number of different symbols that can be sent using a given modulation
- The polynomial Order of a filter transfer function

- Money Order
- Order (business), an instruction from a customer to buy
- Order (exchange), customer"s instruction to a stock broker
- Order of degrees in the Elliott wave principle

- Court Order, made by a judge; a restraining Order, for example, is a type of injunction
- Executive Order, issued by the executive branch of government
- General Order, a published directive from a commander
- Law and Order (politics)
- Order, or military command
- Social Order, referring to the conduct of society
- Standing Order, a general Order of indefinite duration, and similar ongoing rules in a parliament
- World Order, including the concept of a world government

- Architectonic Orders: see classical Order
- Public Order, a concept in urban planning

- Order (decoration), medal or award

- Chivalric Order, established since the 14th century
- Fraternal Order
- Holy Orders, the rite or sacrament in which clergy are ordained
- Military Order, established in the crusades
- Monastic Order, established since circa 300 AD
- Religious Order
- Order (organization), an organization of people united by a common fraternal bond or social aim
- Tariqa or Sufi Order
- Cardassian military unit in the fictional Star Trek universe
- Order of the Mishnah, the name given to a sub-division of the Mishnah, a major religious text
- Order of the Mass is the set of texts of the Roman Catholic Church Latin Rite Mass that are generally invariable.
- The Order (group), an underground American neo-Nazi organization active in 1983 and 1984.

Euler method	In mathematics and computational science, the Euler method is a first-order numerical procedure for solving ordinary differential equations (ODEs) with a given initial value. It is the most basic kind of explicit method for numerical integration for ordinary differential equations.

Chapter 9. NUMERICAL SOLUTIONS OF ORDINARY DIFFERENTIAL EQUATIONS

Chapter 9. NUMERICAL SOLUTIONS OF ORDINARY DIFFERENTIAL EQUATIONS

	Consider the problem of calculating the shape of an unknown curve which starts at a given point and satisfies a given differential equation.
Predictor-corrector method	In mathematics, particularly numerical analysis, a Predictor-corrector method is an algorithm that proceeds in two steps. First, the prediction step calculates a rough approximation of the desired quantity. Second, the corrector step refines the initial approximation using another means.
Runge-Kutta method	In mathematics, the Runge - Kutta method is a technique for the approximate numerical solution of a stochastic differential equation. It is a generalization of the Runge-Kutta method for ordinary differential equations to stochastic differential equations.
	Consider the Itô diffusion X satisfying the following Itô stochastic differential equation $$dX_t = a(X_t)\, dt + b(X_t)\, dW_t,$$ with initial condition $X_0 = x_0$, where W_t stands for the Wiener process, and suppose that we wish to solve this SDE on some interval of time [0, T].
Average	In mathematics, an Average, central tendency of a data set is a measure of the "middle" or "expected" value of the data set. There are many different descriptive statistics that can be chosen as a measurement of the central tendency of the data items. These include mean, the median and the mode.
Taylor's theorem	In calculus, Taylor"s theorem gives a sequence of approximations of a differentiable function around a given point by polynomials (the Taylor polynomials of that function) whose coefficients depend only on the derivatives of the function at that point. The theorem also gives precise estimates on the size of the error in the approximation. The theorem is named after the mathematician Brook Taylor, who stated it in 1712, though the result was first discovered 41 years earlier in 1671 by James Gregory.
Differential equation	A Differential equation is a mathematical equation for an unknown function of one or several variables that relates the values of the function itself and its derivatives of various orders. Differential equations play a prominent role in engineering, physics, economics and other disciplines. Visualization of airflow into a duct modelled using the Navier-Stokes equations, a set of partial Differential equations.
	Differential equations arise in many areas of science and technology: whenever a deterministic relationship involving some continuously changing quantities (modelled by functions) and their rates of change (expressed as derivatives) is known or postulated.
Function	In mathematics, a function is a relation between a given set of elements (the domain) and another set of elements (the codomain), which associates each element in the domain with exactly one element in the codomain. The elements so related can be any kind of thing (words, objects, qualities) but are typically mathematical quantities, such as real numbers.
	There are many ways to represent or visualize functions: a function may be described by a formula, by a plot or graph, by an algorithm that computes it, by arrows between objects, or by a description of its properties.

Chapter 9. NUMERICAL SOLUTIONS OF ORDINARY DIFFERENTIAL EQUATIONS

Chapter 9. NUMERICAL SOLUTIONS OF ORDINARY DIFFERENTIAL EQUATIONS

Multistep methods	Linear multistep methods are used for the numerical solution of ordinary differential equations. Conceptually, a numerical method starts from an initial point and then takes a short step forward in time to find the next solution point. The process continues with subsequent steps to map out the solution.
Unstable	Instability in systems is generally characterized by some of the outputs or internal states growing without bounds. Not all systems that are not stable are unstable; systems can also be marginally stable or exhibit limit cycle behavior.
	In control theory, a system is unstable if any of the roots of its characteristic equation has real part greater than zero.
Initial value problem	In mathematics, in the field of differential equations, an initial value problem is an ordinary differential equation together with specified value, called the initial condition, of the unknown function at a given point in the domain of the solution. In physics or other sciences, modeling a system frequently amounts to solving an initial value problem; in this context, the differential equation is an evolution equation specifying how, given initial conditions, the system will evolve with time.
	An initial value problem is a differential equation $$y'(t) = f(t, y(t)) \quad \text{with} \quad f : \mathbb{R} \times \mathbb{R} \to \mathbb{R}$$ together with a point in the domain of f $$(t_0, y_0) \in \mathbb{R} \times \mathbb{R},$$ called the initial condition.
Ordinary differential equation	In mathematics, an ordinary differential equation (or ordinary differential equation) is a relation that contains functions of only one independent variable, and one or more of its derivatives with respect to that variable.
	A simple example is Newton"s second law of motion, which leads to the differential equation $$m \frac{d^2 x(t)}{dt^2} = F(x(t)),$$ for the motion of a particle of constant mass m. In general, the force F depends upon the position of the particle x(t) at time t, and thus the unknown function x(t) appears on both sides of the differential equation, as is indicated in the notation F(x(t).)
Finite difference	A Finite difference is a mathematical expression of the form f(x + b) − f(x + a). If a Finite difference is divided by b − a, one gets a difference quotient. The approximation of derivatives by Finite differences plays a central role in Finite difference methods for the numerical solution of differential equations, especially boundary value problems.
Approximation	An Approximation (usually represented by the symbol ≈) is an inexact representation of something that is still close enough to be useful. Although Approximation is most often applied to numbers, it is also frequently applied to such things as mathematical functions, shapes, and physical laws.
	Approximations may be used because incomplete information prevents use of exact representations.
Mathematical model	Note: The term model has a different meaning in model theory, a branch of mathematical logic. An artifact which is used to illustrate a mathematical idea is also called a Mathematical model and this usage is the reverse of the sense explained below.
	A Mathematical model uses mathematical language to describe a system.

Chapter 9. NUMERICAL SOLUTIONS OF ORDINARY DIFFERENTIAL EQUATIONS

Chapter 9. NUMERICAL SOLUTIONS OF ORDINARY DIFFERENTIAL EQUATIONS

Difference equation

In mathematics, a recurrence relation is an equation that defines a sequence recursively: each term of the sequence is defined as a function of the preceding terms.

A difference equation is a specific type of recurrence relation.

An example of a recurrence relation is the logistic map:

$$x_{n+1} = rx_n(1 - x_n)$$

Some simply defined recurrence relations can have very complex (chaotic) behaviours, and they are a part of the field of mathematics known as nonlinear analysis.

Interior

In mathematics, the Interior of a set S consists of all points of S that are intuitively "not on the edge of S". A point that is in the Interior of S is an Interior point of S.

The exterior of a set is the Interior of its complement; it consists of the points that are not in the set or its boundary.

The notion of the Interior of a set is a topological concept; it is not defined for all sets, but it is defined for sets that are a subset of a topological space.

Finite-difference methods

In mathematics, finite-difference methods are numerical methods for approximating the solutions to differential equations using finite difference equations to approximate derivatives.

finite-difference methods approximate the solutions to differential equations by replacing derivative expressions with approximately equivalent difference quotients. That is, because the first derivative of a function f is, by definition,

$$f'(a) = \lim_{h \to 0} \frac{f(a+h) - f(a)}{h},$$

then a reasonable approximation for that derivative would be to take

$$f'(a) \approx \frac{f(a+h) - f(a)}{h}$$

for some small value of h.

Maxima

In mathematics, maxima and minima, known collectively as extrema (singular: extremum), are the largest value (maximum) or smallest value (minimum), that a function takes in a point either within a given neighbourhood (local extremum) or on the function domain in its entirety (global extremum).

Throughout, a point refers to an input (x), while a value refers to an output (y): one distinguishing between the maximum value and the point (or points) at which it occurs.

A real-valued function f defined on the real line is said to have a local (or relative) maximum point at the point x^*, if there exists some $\varepsilon > 0$, such that $f(x^*) \geq f(x)$ when $|x - x^*| < \varepsilon$.

Quotient

In mathematics, a Quotient is the result of a division. For example, when dividing 6 by 3, the Quotient is 2, while 6 is called the dividend, and 3 the divisor. The Quotient can also be expressed as the number of times the divisor divides into the dividend.

Shooting method

In numerical analysis, the Shooting method is a method for solving a boundary value problem by reducing it to the solution of an initial value problem. The following exposition may be clarified by this illustration of the Shooting method.

For a boundary value problem of a second-order ordinary differential equation, the method is stated as follows.

CPSIA information can be obtained at www.ICGtesting.com
Printed in the USA
LVOW080707271211
260995LV00001B/121/P